Design Thinking

Design thinking is a ground-breaking problem-solving process which combines logic, intuition, and systematic reasoning to develop long-term solutions to common engineering challenges and to inspire innovation. Serving as an introduction to the concept as well as a reference point, the book is an essential reading for all engineers.

Following a design thinking approach itself to structure its contents, this book is a key introduction to the process, providing case studies to demonstrate the multiple practical uses of the method. Relevant to sectors such as software development, Mobile App Development, sustainability, and Artificial Intelligence, the book has a wide range of applications. The inclusion of a tools section to focus on popular apps and software aids the reader in practically using the design thinking method. It ends by looking forward to the future prospects of design thinking and the innovations which it can inspire.

The book will be of interest to engineers of all professions, including design and management.

Design Thinking

A Forefront Insight

Edited by
Kaushik Kumar and
Muralidhar Kurni

CRC Press is an imprint of the
Taylor & Francis Group, an **informa** business

First edition published 2023
by CRC Press
6000 Broken Sound Parkway NW, Suite 300, Boca Raton, FL 33487–2742

and by CRC Press
4 Park Square, Milton Park, Abingdon, Oxon, OX14 4RN

CRC Press is an imprint of Taylor & Francis Group, LLC

ISBN: 978-1-032-03905-3 (hbk)
ISBN: 978-1-032-03956-5 (pbk)
ISBN: 978-1-003-18992-3 (ebk)

DOI: 10.1201/9781003189923

Typeset in Times
by Apex CoVantage, LLC

Contents

SECTION III Applications

Preface

The editors are pleased to present the book *Design Thinking: A Forefront Insight*. Book title was chosen understanding the current importance of design thinking as well as for having familiarization with one of the most sort-out technologies for industrial and manufacturing world.

Modern problems need modern solutions. Problems are an integral part of our journey as humans. However, there are few among us who can take a concentrated effort to get down to the root of those problems. The usual route taken involves a quick-fix solution, only to have the problem rear its ugly head back at us in due time. Is there a way out? Yes, there is, and the key to it lies in the design thinking philosophy—a mindset that designers have been using to create memorable products at Apple, Nike, and yes, even Google. Design thinking forms a schematic approach to understand and find solution to problems utilizing the same tools traditionally used by developers of commercial and/industrial products, processes, and environments. In mid of twentieth century, the development of creative techniques and new design methods provided the foundation of the concept called design thinking as a secured approach toward problem solving. Authors like John E. Arnold and L. Bruce Archer were the pioneers in discussing about the concept in their books *Creative Engineering* (1959) and *Systematic Method for Designers* (1965), respectively. The technique was designed to provide solutions giving novel functionality and higher levels of performance at a lower production cost which would ultimately increase saleability. Accordingly, the steps used were Empathize (Problem Understanding), Define (Problem Definition), Ideate (Idea Generation), Prototype, and Testing. In the initial phase, these steps were taken one after the other, but the gap between the steps and groups involved in the same started providing illogical or erroneous results which created looping of steps and interaction between groups at different stages. So design thinking is an iterative process that involves understanding the problem, challenging assumptions about it, and identifying multiple strategies and solutions that might resolve it for good. It is essentially a solution-based approach to solving problems.

What's so unique about it that non-designers need to adopt it? Well, it is a problem-solving technique that uses logic, intuition, and systematic reasoning to develop long-term solutions. And if that doesn't convince you, know that products like the iPhone, AirBnB, and Uber are all the results of a bunch of designers applying the tenets of design thinking. While it is mainly used by designers as a technique, its tenets can help anyone to get on course with solving problems creatively. When a problem occurs, you may be tempted to follow the first idea (solution) that pops in your head, but impulsiveness may not get you the desired result. For example, let's say you're looking to create a sustainable lifestyle. The first idea you might get is to stop using plastics, which is good enough. But as you proceed, you may realize that a life that is 100% sans plastic is never truly feasible, which can cause some demotivation. Instead, you could list out multiple ways of living a sustainable life—planting more trees, using public transportation, cooking your meals, and more. This not only keeps you on track but also ensures that you fulfill your goal in a wholesome manner. Exploration is at the heart of design thinking, and it provides for a much better resolution. By pushing your mind to look for multiple solutions, you are exploring various ways in which to best solve your issue. You would be looking at uplifting the level of your solution from being mediocre to worthy.

This strategy isn't limited just to designers. Can we apply design thinking to the "wicked problem" of designing your job, your career, and even your life? The answer is YES. A wicked problem is a big, ambiguous problem that is poorly defined and poorly bounded. That sounds a lot like the problem of finding work you love—that is, the problem of designing your way to the future you want to have. When people ask for help in identifying which career path to pursue, they're often told to identify their passion. However, this is the wrong approach. This is because studies show that only 20% of the population can identify a singular passion. The other 80% of the population is either passionate about many different things, or no one thing rises to the level of "that's what I want to do for the rest of my life". For that 80%, passion isn't something that they have or find, but something that they work into. That is, these people should try something new out, see how it's working, tweak it, and experiment further. And that's what design thinking is all about. Building a future with design thinking means taking an improvisational view of life and moving forward by "wayfinding".

At the core of the book are several application areas where design thinking can be applied, each treated as a complete chapter. This book also provides an introduction to the design thinking approach and mindset, which gives the readers a flavor on the techniques and methods of design thinking. The application chapters provide discussion and reflection that will lead the reader into a deeper understanding of the nature of design thinking. This book also reveals various online tools that you can utilize during the design thinking process. This book will help anyone interested in design to develop their understanding of how designers think and work. Anyone so interested might be a design student, a design researcher or teacher, a manager, or even a designer who still finds their processes mysterious or difficult. Readers should gain from the book some insight into what it means to be a designer, how designers employ creative thinking skills, and what is known about different aspects of design ability and its development from novice students to expert professionals. We are confident that this *Design Thinking* book will serve as an easy-to-understand guide to facilitate anyone to learn and apply the methods and tools to generate innovative ideas and allow to grasp the low-hanging fruits in the short term and design a better system in the long run.

Design thinking can also be used to solve personal problems and to design and build your future. We understand that everyone can learn to master design thinking to become a better problem solver.

The chapters of the book are segregated in *three sections* namely *Section I: Tools*, *Section II: Methodology*, and *Section III: Applications*.

Section I contains Chapter 1, whereas *Section II* contains Chapter 2 to Chapter 7, and *Section III* consists of Chapter 8 to Chapter 11

Section I of the book starts with Chapter 1 explaining the principles of design thinking and its phases, and it also illustrates the various methods and tools available for design thinking. An attempt has also made to illustrate the application of Design Thinking as a process for research and how it can be used in an organizational context. The various steps involved in the process of Machine Learning and Deep Learning are touched upon. At the end, it provides a conclusion that the data collected for any design thinking can be simplified by applying various tools provided in the chapter.

Chapter 2, as the first chapter of *Section II*, provides insight on course design methodology using design thinking. Using an exploratory case study, the use of design thinking as a course design method for designing an undergraduate entrepreneurship course was studied. The most important outcome was the fact that "*despite being most commonly known as product or service design methodology, Design Thinking has shown promise as an innovation-oriented approach to course design*". The chapter elaborates that how unlike traditional course design methods, the design thinking method is more empathetic and focused on the needs of students, is iterative, and supports creative solutions for key problems in the course. Hence, this chapter adds to the limited body of literature on the use of design thinking in higher education for course design and innovation.

The next chapter, i.e., Chapter 3, features design thinking toward product design and product innovation. The design and development of any product possess both engineering and industrial design perspectives to meet the demands and expectations. In this context, design thinking and product innovation tools and techniques have attracted significant attention and play a vital role in the problem-solving approach. This chapter is a noble attempt in providing a basic understanding of design thinking and design innovation using conceptual and exploratory designs by answering the basic four design questions. Further, the detailed discussions over the basic two design approaches of Inside-Out Design Approach (IODA) and Outside-In Design Approach (OIDA) have been elaborated touching the various stages and aspects of Engineering Design (ED) and Industrial Design (ID) during the design and development process of product and innovation.

Chapter 4 discusses design thinking as a modern approach in Artificial Intelligence and Machine Learning. So, this chapter briefs about the amalgamation of AIML (Artificial Intelligence and Machine Learning) and design thinking and how the incorporation of design thinking could bring out the effective changes in the AIML domain.

It is imperative for a human to adapt themselves, and the educational sectors also have to be reinvented. The challenges confronting the educational institutions are the lack of interest and detachment of learners from the learning environment. With this preamble, Chapter 5 briefs the learners' learning using design thinking learning approach (DTLA) with machine learning. A DTLA is one such a method that can be designed by the facilitator for preparing content for learners and employed by the learners for learning and hence to endorse the essential capabilities to prepare learners for the future with a progression in novel teaching and learning methodology.

The next chapter, i.e., Chapter 6, talks about Mobile App Development using design thinking technique. Design thinking plays a vital role in Mobile App Development as it provides an analytical and visualized thought in the designing process. In this chapter, a detailed need of design thinking, the perspective of customer needs, design specifications for apps, applied creativity needed, prototyping of apps, app architectures, and the process of app design thinking are discussed in detail.

The last chapter of the section, Chapter 7, provides a review of design thinking for networking and telecommunications. Institutions like Ericsson and Clearbridge have successfully used the concept for the development of connectivity and smartphone sales, the demand for voice over services, and the initiation of SIM cards. Design thinking provided a complementary approach to the rational problem-solving methods typically emphasized in communication and Network Technology. In this chapter, the improvement in product support using design thinking methodology for Telecommunication and Networking has been critically analyzed discussing Ideation, Prototype, and Testing in telecom software, telecom fiber optic cable market in India, and network scenario for Indian telecom.

The next and last section of the book, *Section III*, handles the application part, and the first chapter of this section, i.e., Chapter 8, talks about the development of laptop workstations to reduce postural risk using the technique. Today, laptops have become one of the important products and have been used by various groups of people from different walks of life. The constant and consistent usage of a laptop without proper posture causes musculoskeletal risk factors and misalignment in the neck and spine curvature. In order to overcome the risk factors associated with improper posture maintained while operating a laptop, an ergonomic laptop stand have been specially designed considering the various posture risks and by adopting the design thinking strategies. The chapter applies rapid upper limb assessment (RULA) on the designed laptop stand, and the results suggested for an application and implementation of the design for a practical environment. The designed laptop stand was then fabricated, and it was found to be satisfactory in minimizing various musculoskeletal risk factors.

In Chapter 9, the authors explored the practical application of using design thinking methodology to design a new medical device. At moments when doctors have to make decisions at short notice, a well-designed product can make the difference between life and death, a fact that influences the motivation of those involved in the development of such important products. The chapter presents step-by-step analysis how a medical device company applied design thinking as a UX and product design methodology in order to develop a new medical device, and, based on the findings, the chapter illustrates the practical benefits and uncovers some challenges when applying design thinking to product design.

In the penultimate chapter of the book, Chapter 10 develops an automatic coin counter and sorter prototype using the concept. Coin sorting and counting are intensive labor works in the developing countries as there is huge flow of coins.

In the chapter, the designing of an automatic sorting and measuring machine has been discussed, which is mainly used for sorting, counting, and packaging a variety of coins. The problem is approached through the systematic steps of design thinking approach, and the prototype has been developed.

The best practice to innovate a product is to learn about the customer desire, expectation, and past experiences with existing product, if any. The last chapter of the section and the book, Chapter 11, describes the process of innovation of a compact and wireless portable food heater to keep the food warm on the go. The product is virtually tested for structural and thermal criteria using an analysis software, and a prototype was developed. In the whole process from inception to execution, the chapter elaborates the usage of design thinking.

First and foremost, we would like to thank God. It was your blessing that provided us the strength to believe in passion, hard work, and pursue dreams. We thank our families for keeping the patience with us for taking yet another challenge which decreases the amount of time we could spend with them. They were our inspiration and motivation. We would like to thank our parents and grandparents for allowing us to follow our ambitions. We would like to thank all the contributing authors as they are the pillars of this structure. We would also like to thank them to have belief in us. We would like to thank all of our colleagues and friends in different parts of the world for sharing ideas in shaping our thoughts. Our efforts will come to a level of satisfaction if the students, researchers, and professionals concerned with all the fields related to design thinking get benefitted.

We owe a huge thanks to each and every Contributing Author, Reviewer, Editorial Advisory Board Member, Book Development Editor, and the team of CRC Press for their availability for work on this huge project. All of their efforts were instrumental in compiling this book, and without their constant and consistent guidance, support, and cooperation we couldn't have reached this milestone.

Last, but definitely not the least, we would like to thank all individuals who had taken time out and helped us during the process of writing this book, and without their support and encouragement we would have probably given up the project.

Kaushik Kumar
Muralidhar Kurni

Editor Biographies

Dr. Kaushik Kumar is Associate Professor at Birla Institute of Technology, Mesra, Ranchi, India. His areas of teaching and research interest are Composites, Optimization, Non-conventional Machining, CAD/CAM, Rapid Prototyping, and Quality Management Systems.

Dr. Muralidhar Kurni is currently working as an Associate Professor in the Department of CSE, at Anantha Lakshmi Institute of Technology & Sciences, Ananthapuramu, India. He is also an Independent Consultant for Pedagogy Refinement, EduRefine, India. He received his Ph.D. in Computer Science and Engineering from JNTUA, Ananthapuram, in 2021. He has more than 22 years of teaching experience. He is the senior member of IEEE and professional member of ACM. His research interests include Learning Analytics, Learning Strategies, Digital Pedagogy, Design Thinking, Pedagogy Refinement, and Engineering Education Research.

Editor Biographies

[illegible] Kaushik Kumar [illegible] Associate Professor [illegible] Birla Institute of Technology, [illegible] Ranchi, India. [illegible] research interest areas [illegible] Non-conventional Machining, CAD/CAM, Rapid Prototyping, and [illegible].

Dr. [illegible] is currently working as an Associate Professor in the Department of [illegible] [illegible] Consultant for [illegible] Computer Science and [illegible] 2021. He has more than 12 years [illegible] of IEEE [illegible] [illegible] Research.

Contributors

Archana A.
PSG College of Technology
Coimbatore, India

D. Magdalene Delighta Angeline
Joginpally B.R. Engineering College
Hyderabad, India

Sudharshana Venkatesh C.
Department of EEE
Thiagarajar College of Engineering
Madurai, India

A. Kanaka Durga
Stanley College of Engineering and Technology for Women
Hyderabad, India

S. Elangovan
Department of Production Engineering
PSG College of Technology
Coimbatore, India

I. Felcia Jerlin
Holy Cross Engineering College
Thoothukudi, India

D. Kavitha
Department of EEE
Thiagarajar College of Engineering
Madurai, India

A. Kishore Kumar
Department of Robotics & Automation
Sri Ramakrishna Engineering College
Coimbatore, India

Kaushik Kumar
Birla Institute of Technology
Mesra, Ranchi, India

R. Manish Kumar
Department of Production Engineering
PSG College of Technology
Coimbatore, India

Ranjan Kumar
Birla Institute of Technology
Mesra, Ranchi, India

S. Pratheesh Kumar
Department of Production Engineering
PSG College of Technology
Coimbatore, India

Brindha M.
PSG College of Technology
Coimbatore, India

Kanmani M.
PSG College of Technology
Coimbatore, India

Sibi Kumar M.
Department of EEE
Thiagarajar College of Engineering
Madurai, India

Siva Subramaniyen M.
Department of EEE
Thiagarajar College of Engineering
Madurai, India

Tharun K. M.
Department of EEE
Thiagarajar College of Engineering
Madurai, India

G. Madhan Mohan
PSG College of Technology
Coimbatore, India

R. Mohanraj
Department of Production Engineering
PSG College of Technology
Coimbatore, India

Anja Svetina Nabergoj
University of Ljubljana
Ljubljana, Slovenia

Anupama Namburu
School of Computer Science Engineering
VIT-AP University
Guntur, India

Mridula G. Narang
M. S. Ramaiah University of Applied Sciences
Bangalore, India

B. Nitish Narayanan
Department of Production Engineering
PSG College of Technology
Coimbatore, India

T. Nivethitha
Department of Electronics and Communications Engineering
Hindustan College of Engineering and Technology
Coimbatore, India

Rajeshwari B. P.
PSG College of Technology
Coimbatore, India

Vladimir Pakrac
University of Ljubljana
Ljubljana, Slovenia
and
MESI Ltd
Ljubljana, Slovenia

P. K. Poonguzhali
Department of Electronics and Communications Engineering
Hindustan College of Engineering and Technology
Coimbatore, India

T. Prabakaran
Joginpally B.R. Engineering College
Hyderabad, India

M. Pranaavh
Department of Production Engineering
PSG College of Technology
Coimbatore, India

R. Naveen Raj
Department of Production Engineering
PSG College of Technology
Coimbatore, India

Ruba Dharshini S.
PSG College of Technology
Coimbatore, India

Senthil Vinod S.
Department of EEE
Thiagarajar College of Engineering
Madurai, India

Supriya M. S.
M. S. Ramaiah University of Applied Sciences
Bangalore, India

Prabha Selvaraj
School of Computer Science Engineering
VIT-AP University
Guntur, India

M. Senthilkumar
Department of Production Engineering
PSG College of Technology
Coimbatore, India

Rok Stritar
University of Ljubljana
Ljubljana, Slovenia

C. Vigneswaran
Department of Production Engineering
PSG College of Technology
Coimbatore, India

Blaž Zupan
University of Ljubljana
Ljubljana, Slovenia

Section I

Tools

1 Tools for Design Thinking

A. Kanaka Durga
Stanley College of Engineering and Technology for Women, Hyderabad, India

CONTENTS

1.1 INTRODUCTION: BACKGROUND AND DRIVING FORCES

The dynamic business scenario involves a continuous change process. The biggest driving force is the application of tools for the better utilization of technological resources whenever we think of a design process.

DOI: 10.1201/9781003189923-2

If we think of the background of design thinking (DT) process, it can be noted that it dates back to use by designers in the past. Later there was a paradigm on the research, and it was felt that it is highly useful to explain various paths of DT. Finally, all these shaped out to a whole distinct field known as design thinking today.

The uniqueness of design thinking is its non-linear iterative process and the various teams working on design thinking focusing on understanding users, the challenging assumptions they make, and then arriving at prototype and testing [1]. Generally, when the problems are not defined properly or defined incorrectly, the significance of design thinking is highlighted. This is achieved by ideating the problem several times, reframing the correct ones using techniques like brain storming and so on and then trying to prototype and test the model/design. It is said that design thinking is a human centred approach to innovation that draws from the designer's tool kit to integrate the needs of people, the possibilities of technology and the requirements for business success. We can notice that there are two broad types of design thinking, namely problem based and solution-based thinking. While solution-based thinking emphasises on achieving the solutions to tackle the problems effectively, problem-based thinking focuses on the hindrances and how to fix them, considering the limitations.

1.1.1 The Four Principles of Design Thinking

- **The human rule:** This focuses on human-centric approach of design thinking. Without any reference to the context, all design activity is social in nature, and any social innovation will lead to human-centric approach.
- **The ambiguity rule:** However, the best is the design thinking model, which leaves a space for ambiguity, and we cannot remove or confine this to zero ambiguity. This rule helps us to see things differently to narrow down our limits of knowledge and ability.
- **The redesign rule:** It is said that we need to learn, unlearn, and relearn. In spite of technological changes, basic human wants remain unaltered. We shall only redesign so that the expected outcomes are achieved.
- **The tangibility rule:** This rule emphasizes that the design should be visible and tangible or perceived. Tangibility is achieved with the use of prototypes so that designers can communicate them very effectively.

1.1.2 Methodology

The various types of design thinking on the data collected are studied, and the same are anchored with the relevant tools for design thinking at the relevant phases. All the tools applicable for various stages of design thinking have been provided, with a clear segregation of and application to the particular phase of DT.

1.1.3 Design Thinking Comprises Five Phases

1.1.3.1 Phase 1: Empathize

This is the first stage of design thinking. In this stage of design thinking, we need to spend a lot of time in understanding the users and their needs and objectives. At this stage, it is essential to get engaged with people to know their psychological and emotional levels, keeping aside the assumptions of the designer so that real insights are obtained.

Online tools to support Empathizing are as follows:

- For collecting (raw) information use *Zoom, Typeform.*
- *Creatlr* is an application where all designers can be on the same page and organize their thinking.

1.1.3.2 Phase 2: Define

It is the next stage after Empathize.

All the findings collected in the Empathize stage are reviewed critically to fix and define the problem. It is essential to critically review the problem from the user-centric point of view.

1.1.3.3 Phase 3: Ideate

As we have clear insights from the Empathize and Define stages, we shall try to work on the potential solutions for the problem at this stage. This is a stage where a lot of creativity is involved. Other techniques are brainstorming, mind mapping, and role play (bodystorming). Here, the designer shall try to challenge the established beliefs, which may at times prove to be wrong. At the end, at the end of the stage, we should narrow down the ideas from large to very few and progress further.

1.1.3.4 Phase 4: Prototype

The fourth stage is where a few selected ideas from the previous stage are experimented to see if we get the visibility of the products. It is basically an improvised model of the product that considers the solutions of the earlier stages. This is a very vital step in cross verifying each solution so that any constraints and flaws are captured and eliminated. This can be further improvised or redesigned or rejected depending on the confidence of the outcome and to achieve the final prototype.

1.1.3.5 Phase 5: Test

This is the final stage, which follows prototype stage. We cannot conclude that Test is an end of DT, as at times, after the Test of the prototype, we may repeat all the previous stages again, until an amicable and proper solution from the prototype is received, if required, to redefine the original problem.

1.1.4 Design-Thinking Methods and Tools [2]

The most widely used tools are:

- **Persona**: It helps to find users' expectations and wants.
- **Stakeholder map:** It is a visual map of all the members of the group for a given product or service.
- **Customer journey map (CJM):** It takes us through the entire stage of movement of the customer right from ordering till delivery and payment.
- **Service blueprint**: It is a tool that defines the roles of stakeholders who are involved in customer service.
- **Rapid prototype (RP)**: It is prototype emphasizing on Customer Experience Management.

1.1.5 Tools that Strengthen Defining Stage of DT

- **Smaply:** It is a tool giving a detailed description of the people involved in the process which is editable.
- **Userforge**: This tool helps to minimize the number of clicks about the people, which is taken in the software.
- **MakeMyPersona:** It is a tool that gives complete information about the buyers. It is navigated through 19 questions about the buyers.

1.1.6 Ideation Tools

- **Whiteboard Apps**: They are appropriate for organizing ideas and prioritizing.
- **Stormboard**: Basically a tool with reporting choices.
- **Mural, Miro**: It emphasizes on visualization on workflow.
- **Ideaflip:** A simple yet elegant tool for brainstorming sessions either with your team or alone. Anyone can add their ideas on it or post on it like notes to the virtual space.
- **Ideaflip**: It gives various options of ideas which can be flipped according to suitability.

1.1.7 Prototyping Apps

- **Boords** is a tool involving pictures and GIFs (graphics interchange formats) for storytelling.
- **Mockingbird** is a prototype tool for testing beforehand.
- **POP** involves in combining the sketches and pictures.

1.1.8 Testing Applications

- **UserTesting.com** [3] It explains the ease of testing a design for given set of users.

- **Hotjar** involves feedback based on polls and surveys.
- **Pingpong** is useful for scheduling interviews for subsequent analysis.

1.1.9 Design Thinking Tools for a Full-Fledged Process [4]

The following are the design thinking tools for a full-fledged process.

- **Sprintbase:** It is an engine used to solve the problems virtually.
- **InVision:** It is an app that provides solution-based frame work using several ideas and potential solution from these various ideas.
- **Collaborative online boards**: *Mural* and *Miro* (earlier known as Real-time Board).

1.1.10 Other Tools

The following are the other tools.

- **Innovation flowchart**—It allows to map out the process before it starts.
- **Question ladder—**This worksheet is helpful to define your questions and disperse workload.
- **Design thinking tool kit**—It is available in simple language and more user-friendly mode.
- **IDEO design kit**—This is mostly useful for social organizations and non-government organizations. This is an extremely useful tool for beginners on design thinking.
- **Google ventures design sprint**—It is very useful for small teams to tackle quickly, taking less than 5 to 6 days.
- **Design thinking mix tapes**—There are different types for each stage of the process, which is widely useful for keeping the process under control.

1.2 CASE STUDY

Out of the various applications we can have for design thinking, we bring about the five case studies as follows:

1. *Design thinking for research*
2. *Implementation of design thinking in a company*
3. *Machine Learning and deep learning*

Let us take the case study of application of design thinking for research applications.

1.2.1 Case Study: Design Thinking for Research

Design thinking for research can be carried out using *immersion tools*, general research, interviewing the people, taking general sessions of the various stake holders. The tools like Inside Cards, Affinity Diagram and Concept Maps, Empathy Map, Personas, and Blue Print can also be used.

Following tools are predominantly used for design thinking for research:

- ***Reframing tool:*** Unanswered questions in a company can be attended using this tool which helps to eliminate biases and assumptions about a business, product, or service.
- ***Ideation tools*** depend on the concepts like Brainstorming, Co-creation Workshop, Idea Menu, Positioning Matrix, and Brainwriting.
- ***Prototyping tools*** involve Prototype on Paper, Volume Model, Staging, Storyboard, and Service Prototype.

1.2.2 Implementation of Design Thinking in a Company [5]

Design thinking needs a specific attention and cannot be developed over a very short period of time.

Steps to implement are as follows:

- **Concentrating on the problem**
  Most of the times, in organizations, failures to solve problems are due to not being able to correctly identify the problems from the start of the project. Following are the tips to overcome this defect where we can identify the problems from the beginning:
  - **Listen**: Getting firsthand information from users.
  - **Seek:** Seeking questions from the people facing the problem and also finding what happened to their earlier efforts to curtail this.
  - **Have collaborative conversations:** Never entertaining working in silos – encouraging participative management so that everyone can contribute.
  - **Stay impartial:** Not having any bias in quantifying the problem and solution.
- **Imparting design thinking skills for the team**
  The design thinking process is not handled by one department or function, though it is handled by project managers or engineers. As design thinking involves getting answers from questions, so that it can be understood properly and tested, it emphasizes on participation from everyone.

Steps to develop design skills for the team are as follows:

- **Inculcating the DT mindset:** When it gets complicated with the involvement of too many people, we should find new ways of modifying this process using surveys and common consensus. New results shall be welcomed.
- **Promoting interests in design thinking:** Motivating people who can contribute with their new ideas and help them expanding their skill set.

  - **Asking more questions:**
    As design thinking is an iterative process, we shall seek questions from the users repeatedly.
- **Developing a good learning and feedback process:**
  - We should focus on things that were not properly handled.
  - We should correct the failures for future success.
- **Embracing the feedback loop**
  As we may not get the best solution in the first cut, we shall keep trying all the possible alternatives to get the best one. This can be achieved through a proper feedback loop mechanism.

Tips for feedback process are as follows:

- **Iterative testing**
- **Conduct feedback sessions frequently**
  Many of the corporate companies have innovative teams consisting of consultants and hired specialists for DT—in view of the great importance of design thinking. Some of the resources are *Join.me, Ethnio,* and *Lookback*. Design thinking is not industry-specific and can be applied to any problem and help streamline a business process with simplicity.
- **Immersion**
  This is the first tool that is utilized in any process of design thinking. This is an idea for a product that can change life, which, as we come to think of, that is, immersion begins to happen. Some of the design thinking tools for Immersion are: *SessionLab, Stormboard, IdeaFlip, Smaply, Userforge, MakeMyPersona.*

Immersion includes various sub-processes:

  - ***Simulation:*** Creating the real-world environment pertaining to the problem.
  - ***Research:*** After simulation, exploratory research related to the context surrounding the problem begins. Again, it can be multifaceted depending on the data that is to be collected and used.
  - ***Brainstorming:*** This tool is most crucial when it comes to collaborative brainstorming sessions with a team, stakeholders, or prospective innovators/product manager.
  - ***Overshadowing:*** This is the fourth phase of ideation, which is used to monitor a client's interaction with a product over a considerable period to understand preferences and dislikes.
- ***Visualization*:** It uses any process that involves human invention from industrial manufacturing to scientific discovery.
  - In design thinking, the visualization tool is crucial. This is the second step after Step 1(Immersion), which helps to visualize how to solve the problem.
  - Visual thinking helps designers to imagine/create a niche product that is suitable for the needs of the market it wishes to bring about a change in.

    - Tools available in the market for *Visualization* are: *Miro, Conceptboard, Google Jamboard, Mural, and Shape.*
- ***Mind Mapping***
    - It helps to generate, structure, interlink, and classify ideas to look for patterns for the final design.
    - An intermediate phase overlaps through all design thinking processes.
    - Mind Mapping can happen only in a team setup. Here the *customer feedback on patterns* are used to present to them.
    - In case of products for niche markets, we should retrieve feedback on patterns for developing a product with zero loopholes or an assembly line of products.
    - **For example,** though Tesla's Electric Car is not an innovation, the use of a design thinking tool—mind mapping—helped them identify the challenges around electric cars to now have the most the successful and profitable EV fleet in the world.

Design thinking tools for mind mapping [6] are *ClickUp, Ayoa, Miro, SmartDraw, MindMeister, Milanote, Microsoft Visio, MindGenius, and Lucidchart.*

- **Rapid Iteration**
    - The designers from the company also apply the Rapid Iteration tool to test designs quickly in the real world.
    - For example, in the case of e-commerce platforms or food delivery organizations like Swiggy, all stakes are time bound.
    - Testing designs quickly in the real world, improving upon them, and retesting keeps the market open for innovation.
    - If you have spent too much time solving small logistical problems, the entire chain of operations can be affected.
- **Assumption Testing**
    - Assumption Testing is an important design thinking tool.
    - For example: In 2009, two Harvard Business School graduates launched Rent the Runway on the assumption that customers like to rent dresses over the Internet.
    - Assumption testing implies identifying assumptions, stereotypes, and prejudices that underlie the attractiveness of any new product and then testing it. It involves thought testing and field experiments.
    - This design thinking tool uses large amounts of data to classify
        - what might work,
        - what generally works, and
        - what should not work and why it works.

    Example: The case of Fair & Lovely, which is now Glow & Lovely.
- **Prototyping**

    This is the tool that turns the ideas into material reality.

- Prototyping is about minimizing the I of ROI (Return on Investment) to the extent that the cost of your first 2D prototype is reduced to that of a pencil or a piece of bread.
- Most businesses these days proceed with 2D prototypes before going on with 3D models. Therefore, prototypes can also include storyboards, images, role-play, skits, etc.

Tools available for Prototyping are *Boords, Mockingbird, POP-app by Marvel, PowerPoint, Keynote, Lumen5, or Moovly*, and *Proto.io.*

- **Finding the Value Proposition**
  - Rather than adding more features to the product, we should focus on finding value. However minimal that may be, the value should be irreplaceable.
  - For example, *Apple.* After Steve Jobs left the organization in 1985, they could not succeed in its product development cycle. Only when a new line of products using design thinking were introduced by Steve Jobs again, in no time, the company became a market leader in smartphones and computers.
  - There are no specific tools in the market for creating a roadmap of the value proposition for the stakeholders or clients, but it suffices to say that some basic ones like *Strategyzer's Value Proposition Canvas, emaze, Canva* [7], and *PiktoChart* [8] will do the job.
- Learning Launch

  Once your prototype is approved, it goes to the final end user, your customer. Learning Launch is the tool wherein your test product is launched in the market for a quick experiment.
  - In contrast to a new product launch, this test is conducted solely for gathering data.
  - The Learning Launch process is based on the Darden Growth Leader Research Project.
  - Since there is no perfect algorithm to achieve the perfect product, design thinking uses this tool to learn along the way.
  - For example, post-lockdown, the French government's initiative to make the country green, sustainable, and inclusive translated into making 650-km bicycle lanes throughout the territory. It was an urban planning project that worked along the learning launch process to see what works best.
- **Collaborative Creation**
  - A crucial component of the testing phase is also data analysis using the feedback of multiple stakeholders thus making value-added differentiation possible in a market of infinite products and services.

- A tip that always works in this stage is to present the customer with an incomplete product. This makes intuition work and is a good way to engage a buyer's creativity and intellect. Offering multiple prototypes can make this process even better for zeroing in on your final product.
- Tools available in the market for helping co-create are *Excel, Tableau, Xplenty*, and *Digsite*. One can even use *UserTesting* for real-time feedback.

1.2.3 Machine Learning and Deep Learning

1.2.3.1 Step 1: Empathize and Analyze

This is the first stage of design thinking. Here, the importance is given to understanding the problem and to solve it.

1.2.3.1.1 Design Thinking

In this stage, we involve the users and experts for a collaborative understanding of the problem.

1.2.3.1.2 Deep Learning

Here, neural network is used to understand the real-world problem which involves taking decisions and arriving at the variables and metrics to address it.

1.2.3.2 Step 2: Define and Synthesize

For ensuring the relationships, data is analyzed by splitting into training set and testing set.

1.2.3.3 Step 3: Ideate

Design stage is ideated for evaluating suitable alternatives.

1.2.3.4 Step 4: Prototyping

The training data is modified to suit the requirements of results.

1.2.3.5 Step 5: Test and Validate

This can be summarized into three major steps:

After testing the data set, it is validated for end-user requirements and fine tuning it further.

1. **The workflow for AI projects:** AI projects can be tested based on the standard IBM procedures as follows:
   - Data collection
   - Exploratory data analysis
   - Data visualization

2. Missing data can be compiled for the following:
   - MCAR (Missing Completely At Random)
   - MAR (Missing At Random)
   - MNAR (Missing Not At Random)
3. We can recheck the aforementioned missing data using *Univariate Imputation, Multivariate Imputation, Transformation, and Modelling/ Testing.*

Benefits of design thinking at work are the following:

- *Significantly reduces time-to-market*
- *Helps in cost reduction and is a great Return On Investment*
- Provides *enhanced customer retention and customer loyalty*
- *Contributes for innovation*
- *Can be applied across the company*

1.3 CONCLUSION

From the study of this chapter, one can find the various types of design thinking applications available. It can also be noted that the process of data analysis for design thinking is much simplified by the usage of relevant design thinking for the respective stage of design thinking. This will be of much use and saves a lot of time and energy while the data collected is studied for a particular design thinking as one can directly use the relevant tool(s) for the design thinking of the relevant phase. Three important case studies, viz., Design Thinking for Research, Design Thinking Implementation for your Organization, and Design Thinking for Machine Learning and Deep Learning have been elaborated with the relevant tools applicable. At the end, the benefits of design thinking at work are explained.

REFERENCES

[1] *Design Thinking for Dummies*—Christian Muller. Roterberg: Wiley.
[2] Design Thinking Methods and Tools for Innovation—https://springer.com [Accessed on Dec 21, 2021].
[3] UserTesting.com [Accessed on Dec 21, 2021].
[4] *The Design Thinking Toolbox: A Guide to Mastering the Most Popular and Valuable* Innovation Methods—Larry J. Leifer, Michael Lewrick, and Patrick Link. Hoboken, New Jersey: John Wiley & Sons, Inc.
[5] *Designing for Growth: A Design Thinking Tool Kit for Managers*—Jeanne Liedtka and Tim Ogilvie. New York: Colombia University Press.
[6] www.toptal.com/designers/digital-product-design/ [Accessed on Dec 21, 2021].
[7] Canva: www.canva.com [Accessed on Dec 21, 2021].
[8] Piktochart: https://piktochart.com [Accessed on Dec 21, 2021].

Section II

Methodology

2 Design Thinking as a Course Design Methodology

Blaž Zupan, Rok Stritar, and
Anja Svetina Nabergoj
University of Ljubljana, Ljubljana, Slovenia

CONTENTS

2.1 INTRODUCTION

In the late 1980s, the first graduate and undergraduate courses in entrepreneurship were offered as being part of an accredited curriculum at the University of Ljubljana, School of Economics and Business (SEB LU). These courses, aimed at teaching students the basic approaches to business planning, were based on a textbook by Timmons and Spinelli (Timmons & Spinelli, 1994). Over the last decade, SEB LU has reexamined the content and structure of their entrepreneurship courses in response to findings from the internal evaluation study that showed that the program was not yet effectively building a culture of entrepreneurship among its students (Drnovšek & Glas, 2002).

The primary goal of the first entrepreneurship classes at SEB LU was to address the need for the emergence of small- and medium-sized private companies in order to support the young emerging market economy in Slovenia after the country got independence in 1991. The courses followed a similar business model-guided design that was well in line with the teaching of entrepreneurship

DOI: 10.1201/9781003189923-4

in other universities worldwide (Honig & Karlsson, 2004). The undergraduate business-model-based course Introduction to Entrepreneurship was found to have several positive attributes, including: (1) stimulating students to consolidate and build on knowledge from past business courses, (2) promoting teamwork, (3) thinking creatively, (4) teaching presentation skills, and (5) teaching students to solve complex business problems using a holistic analytical approach (Stritar & Drnovšek, 2006). Students and faculty members responded favorably to the class, assessing it positively. However, throughout the years, several weaknesses and challenges of focusing on teaching the business models had emerged. A senior faculty member stated:

> *At the beginning there was a lot of enthusiasm and the business plans that students presented were interesting. However, throughout the years we found out that hardly any business ideas were put into practice, the innovativeness of the ideas was decreasing, and students were more focused on perfecting the technical side of the documents instead on developing their business ideas.*

In consideration of the critique, SEB LU iteratively redesigned an undergraduate course for third year students entitled Entrepreneurial Project using a design thinking framework. Like the other SEB LU entrepreneurship courses, the course, which enrolls approximately 80 students per year, originally had a business-plan-guided design. The teaching staff involved in the Entrepreneurial Project course, which includes the authors of this chapter, saw a need to redesign the course to improve students' entrepreneurial skills with a stronger focus on applied business practices.

Although design thinking has been used and discussed by educators more recently, there is very little literature on the use of design thinking in higher education for course design and innovation (Rutherford, 2020). To expand on our understanding of this unexplored area, this chapter presents a qualitative case study of the design thinking approach used to improve an undergraduate course at SEB LU. The purpose of the case study is to explore how the course evolved using this approach for iteration and to analyze the benefits and challenges of using design thinking for course design.

2.2 LITERATURE REVIEW

From the early 2000s, teaching entrepreneurship at higher educational institutions has been receiving reassessment and numerous calls for a change of paradigm, values, and ways of doing things (Gibb, 2002; Kuratko, 2005; Mwantimwa, 2019). Specifically, scholars have found that: (1) contents, which are delivered to students, are seldom supportive of what entrepreneurs actually need to know and do (Collins et al., 2004); (2) educators should be teaching contents through practice rather than solely through theory (Mahmoud-Jouini et al., 2019; Neck & Greene, 2011); (3) programs do not have the intended effect and sometimes even lower students' entrepreneurial intentions (Oosterbeek et al., 2010; Piperopoulos, 2012); (4)

programs do not improve students' cognitive entrepreneurial skills (Huber et al., 2012); and (5) entrepreneurship courses have no impact on graduates' new venture performance (Chrisman et al., 2012; O'Brien & Hamburg, 2019). Some entrepreneurship education still fails to recognize the role of teams (Laukkanen, 2000) and multidisciplinary approaches (Pretorius, 2008), does not sufficiently emphasize developing right-brain creative capabilities (Kirby, 2004), uses obsolete methods (Hytti & O'Gorman, 2004) which teach about the theory of entrepreneurship rather than experiencing practical entrepreneurial experiences, and often fails to recognize the individuality of each entrepreneur (Kirby, 2004).

Since the beginning of university-level entrepreneurship programs in Slovenia and other Eastern European countries, hardly any changes in how to teach entrepreneurship have been made, which, as some have argued, has resulted in outdated educational programs, inadequately educated and inexperienced faculty, lack of theoretical and practical understanding of what entrepreneurs need, and, consequently, the absence of any measurable impact of entrepreneurship education curricula on students' entrepreneurial performance (Bae et al., 2014; Fayolle & Gailly, 2009).

Also, research on entrepreneurship education in higher education institutions in Slovenia and other post-communist countries concluded both that universities did not adequately train teachers of entrepreneurship and that there was insufficient use of real entrepreneurs in teaching programs (Luka, 2019; Varblane & Mets, 2010). Another study of entrepreneurship education concluded that Slovenian universities need to pay more attention to the fulfillment of students' expectations, the quality of teaching materials, and the selection of the lecturer who should be able to teach in an interesting and clear way (Antoncic et al., 2005). As the entrepreneurship courses were apparently outdated, there was a strong need to redesign the courses.

Course design is methodologically well-covered, with different goals and techniques, and has been evolving as a field for over 60 years (Tyler, 1950). Some common course design techniques include an assignment-centered approach where instructors list expectations, identify assignments, create a course schedule, and develop activities to complete the assignments (Walvoord & Anderson, 2009); an objectives-based approach, which includes determining the backgrounds and interests of the students likely to enroll, the learning objectives, the scope and content of the course, and an evaluation component (Davidson et al., 1994; Zheng, 2018); and a conceptual approach, which includes designing a conceptual map of topics, making clusters of topics, deciding on a sequence of subject matter in each cluster, and determining the number of problems and design problems covering the topic (Bjorklund et al., 2019; Posner & Rudnitsky, 1994). These approaches are generally top-down in terms of processing, as teaching staff chooses the learning objectives and activities to achieve learning objectives based on their own training and opinions of educational theory and pedagogy, rather than them being based on students' expressed needs and ideas.

Design thinking has some promise as an innovation-oriented approach to course design that fosters learning rather than teaching (Hong & Sullivan, 2009;

Loughran, 2013; Whetten, 2007). Educators have identified general design skills as a valuable tool in course design (Falvo & Urban, 2007) and have come to the realization that designers in all fields, including course design, use very similar methods (Hokanson et al., 2008). Some of the course design models described previously have been labeled as being somewhat outdated as they do not rely intensively on understanding and identifying the actual needs of students, which in this case are the "consumers" (Visscher-Voerman & Gustafson, 2004). Some scholars have said that traditional course design methods lack a holistic approach (Van Merriënboer & Kirschner, 2013), neglect the importance of prototyping (Tripp & Bichelmeyer, 1990), neglect the effect of diverse teaching—learning situations and teams (Booyse, 2010), and focus too great of an extent on the use of textbooks and transfer of obsolete knowledge (Shawer, 2010).

Design thinking generally can be defined as an approach to creative problem solving that draws on the tools and skills designers apply to problem solving (Schweitzer et al., 2016). Design thinking method started getting used in businesses, where numerous organizations used it to increase sales and profitability (Ward et al., 2009), and in academia when some of the leading design, engineering, and business schools adopted it as a teaching method. More recently, design thinking has been used by educators to improve school layouts, teaching (Cankar, Deutsch, Zupan, & Setnikar Cankar, 2013), and course design (IDEO, 2014).

Design thinking is typically described as a cyclical process that has an established sequence of steps to be followed and is repeated several times in order to ensure a viable solution (Rauth et al., 2010). Figure 2.1 illustrates the steps in the design thinking process described by Rauth et al. (2010). The first step, "Empathy", includes exploring the problem space from a human-centered perspective, with a focus on listening to the experience and needs of consumers and involved stakeholders. The next steps are "Define", which includes articulating the problem or problems; "Ideate", which includes brainstorming solutions and selecting solutions to further develop; "Prototype", which includes developing a prototype, or series of prototypes, of all or part of the proposed solution; and "Test", which includes implementing and testing the prototype through constructive feedback from users. In general, the steps are followed linearly, but going back, rethinking, and reiterating are highly encouraged to improve the quality of the prototype or final solution or simply to gain a different perspective on the problem.

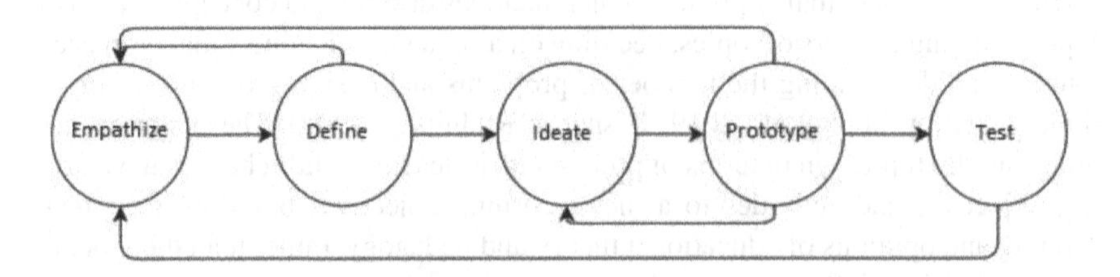

FIGURE 2.1 The design thinking process (adapted from Stanford University, 2007).

Several mindsets have also been identified as an important part of design thinking methodology (Brown, 2008; Fraser, 2007; Nussbaum, 2004; Rauth et al., 2010). In particular, design thinking:

- is human-centered: people are the source of inspiration and focus of problem solving;
- is mindful of process: design thinkers employ an iterative methodology to explore numerous possible solutions and learn from failures;
- is empathetic: to successfully solve an individual's problem, that individual's feelings, thoughts, and attitudes must observed, experienced, and understood;
- includes storytelling: an important tool with which to communicate observed user needs;
- has a culture of prototyping: the process is experimental and iterative, builds on experience, and tests intermediate solutions;
- is biased toward action: all skills and tools should be practiced;
- includes radical open-minded collaboration among disciplines (Higgins et al., 1989): multidisciplinary teams will produce better results if Design Thinkers have the ability to build on the ideas of others;
- includes integrative thinking: using abductive reasoning (Martin, 2007) dramatically improves existing products;
- is optimistic: establishing there is always a solution; and
- challenges constraints and supports creative solutions: obstacles and constraints need to be challenged for creative and sometimes highly unorthodox solutions to succeed.

While design thinking has become the part of popular vocabulary in teaching practice as well as in business and management, various other settings might benefit from the application of its principles as well. The premise is that by knowing about the process and the methods that designers use, educators will be better able to take the effectiveness of course design to a higher level.

2.3 METHODOLOGY

To examine the benefits and challenges of redesigning the course using the design thinking approach, we used an exploratory, qualitative case study methodology (Lea, 2004). Case studies are often used to document the procedures of a particular event so as to enrich the thinking and discourse regarding the development of educational theory by systematic and reflective documentation of experience (Stenhouse, 1988; Yin, 2003). We choose such a design as an exploratory case study because it is more likely to fit with reality than the one formed by combining insights from prior literature (Eisenhardt, 1989), and it allowed us to search for fresh perspective and insights in a traditional research field (Mäkelä & Turcan, 2006). The aim of our approach was to produce new concepts and hypotheses that can later be tested with deductive methods (Eisenhardt & Graebner, 2007). The research questions of the case study were:

1) How was the design thinking approach applied to redesign the course?
2) What key findings emerged from the design thinking approach that prompted changes to the course?
3) What were the key benefits of the design thinking approach to course (re)design?
4) What were the challenges of the design thinking approach to course (re) design?
5) How did the design thinking approach compare to traditional methods of course design?

To gain a rich insight into the process of evolving curriculum design and to map the process, we gathered data from two sources: first was the interviews with faculty members involved in designing and teaching the course, and second was the review of class documentation including syllabi and class materials. Semi-structured interviews were conducted with two senior and four junior faculty involved in the course. The interviews were approximately 1.5 hours in length and were facilitated using open-ended questions focusing on the faculty's experiences with the course redesign, how and why changes to the course were made, and the results of those changes. Using a grounded theory approach (Strauss & Corbin, 1997), the interviews were coded by two raters. The analysis began with five major codes used to identify excerpts related to the five main components of the design thinking process applied to the course redesign: empathy, problem framing, ideation, prototyping, and testing. Other codes produced over the course of the analysis, which categorized details about specific influences and changes made to the course, included: mentors from the business world, real projects, skill training—empathy, skill training—sales, business aspects of the student project, individual projects, group projects, field projects, and Startup Weekend (a model of a 54-hour business plan competition that was used as one of the class session formats for the course).

Course materials were also carefully reviewed to identify and document changes in the course syllabus and activities. Two raters reviewed the course materials to summarize their contents, describe the course environment (e.g., structured or unstructured), and identify the new components of the course added each year. Table 2.1 provides a summary of this analysis.

There are some limitations to the case study methodology, particularly the inability to generalize the results to other experiences in course redesign. The single descriptive case study approach is analogous to experiments performed in the physical sciences, where statistical representativeness on a certain population is not claimed, and yet a contribution to a general theory of the phenomenon is sought (Yin, 2003). Although some scholars argue that case studies cannot provide reliable information about the broader context of a phenomenon (Dogan & Pelassy, 1984; Platt, 1992), our research aligns with the argument that case study methods can produce rich insights and can contribute to a new and cumulative body of knowledge (Flyvbjerg, 2006; Meyer, 2001). In this case, the analysis of the design thinking process applied to one course at SEB LU helps us consider the design thinking process as a methodology for course design in higher education and create new theories for further testing.

2.4 RESULTS

Results of the interviews and content analysis are presented here under subheadings representing the components of the design thinking process: empathize, define, ideate, prototype, and test. Table 2.1 displays the results of the course

TABLE 2.1
Results of the Content Analysis of the Course Materials (Columns B–D) and Key Benefits, Key Problems, and Lessons Learned from Each Year, Derived from the Faculty Interviews (Columns E–G)

Year 1
Course content & structure
Teams of 3-5 students (self-selected) worked on 4 different long-term entrepreneurship projects prepared by the teaching team (an e-paper, kiosk, moneta, and housing for young families). No structured content or workshops dedicated to prototyping. Emphasis on individual prototyping outside the classroom. Mentors from business and academia were included to help advise students.
New components
Prototyping, hands-on projects, no structured content, and mentors from business and academia to provide students a more hands-on learning experience.
Student feedback tools
Observations, individual consultations, and a formal survey where students assess the quality of the course. The survey included some general questions about the course, the faculty involved with the course and open-ended questions such as, "How satisfied were you with the course?"
Key benefits
Action-oriented with several hands-on experiences. Focus on developing prototypes solving true customer problems. The larger teaching team offered a broader range of experience and knowledge on different subjects.
Key problems
The progress of the entrepreneurial teams varied significantly. Organizational issues arose due to the unstructured course organization. Lack of written textbooks was disorienting for students. Too much focus on solving problems of existing ventures instead identifying new opportunities. No workshops for physical prototyping.
Lessons learned
There was a need for a dedicated prototyping space, a simplified course structure and more diverse student teams.

(Continued)

TABLE 2.1 (CONTINUED)

Year 2

Course content & structure

More structured approach with 10 lectures on the entrepreneurial process, written reports, homework and presentations. Teams of 3-5 students were organized to mix students from different disciplines. Teams worked on long-term entrepreneurship projects (a sailing school, an Argo B2B system development, a USB battery charger). Workshops were added with emphasis on physical prototyping.

New components

Structured content to provide students a more predictable course framework; recruitment of students from different disciplines to create more diverse teams; workshops were added to allow in-class time for prototyping.

Student feedback tools

Observations, individual consultations, and formal survey where students assess the quality of the course.

Key benefits

Multidisciplinary teams of students made the teams more effective. More structured approach and use of workshops resulted in better prototypes.

Key problems

Emphasis on prototype development was too focused on research and development, rather than identifying a market demand. Too much time for the project development meant little progress in each week of the course. The fixed structure of the course had sometimes interfered with the creative processes.

Lessons learned

There was a need for more teaching on identifying market opportunities and more time for creative work on prototyping.

Year 3

Course content & structure

Similar structure as in year 2, but fewer lectures and more emphasis on individual and team work with 4 written homework assignments. More teaching emphasis on identifying market opportunities. Students studied different phases of the design proccss individually.

New components

Emphasis on individual ideas, and more time for individual and group projects to allow more time for learning about identifying market opportunities and prototyping.

Student feedback tools

Observations, individual consultations, and formal survey where students assess the quality of the course.

Key benefits

More time for project development and individual work achieved better project results. Students appreciated the emphasis on individual entrepreneurial ideas.

Key problems

The projects and individual work was not as interdisciplinary. Teaching the phases of design without student interaction with a real-world application was not effective for learning.

Lessons learned

There was a need for more applied learning of the design process.

Year 4

Course content & structure

The course was composed of two projects. The first one was a group project: teams were given a design brief to develop a solution for a student cafeteria. The second part was focused on individual entrepreneurship projects.

New components

Students were taught a design thinking methodology
which they applied to the individual and team projects.

Student feedback tools

Observations, individual consultations, and a formal survey where students assess the quality of the course.

Key benefits

Students responded well to the combination of the design brief to develop students' skills on opportunity identification, and individual projects, which fostered their entrepreneurial potential.

Key problems

While the results of the first project were very good, the individual projects failed to meet expectations of the teaching team, as the overall quality of the individual projects was poor.

Lessons learned

There was a need to improve the learning experience and quality of the individual project. There was too much emphasis on prototyping and ideation and not enough on understanding the customer.

Year 5

Course content & structure

The same structure with two projects was used but with more focus on teaching about understanding customer needs (empathy). Field work exercises were added on customer interaction, such as customer interviews and surveys.

New components

Lectures in empathy, observation, and anthropological methods, with a goal of increasing the quality of the projects.

Student feedback tools

Observations, individual consultations, and a formal survey where students assess the quality of the course. Short survey with open questions administered by the teaching team.

Key benefits

The emphasis on field work and understanding the customer was useful to students.

(Continued)

TABLE 2.1 (CONTINUED)

Key problems

Solutions proved to be poorly thought-out from the business perspective. Some specific entrepreneurship skills were missing from the course training, namely sales.

Lessons learned

There was a need to focus more on business models that conceive a solution that is not only desirable and feasible, but also viable.

Year 6

Course content & structure

The course was divided into two parts: 1) a more traditionally designed course on entrepreneurial opportunity development with the focus on business models and 2) project work. Students had to work on three projects. The first one was 3 Euro Challenge: each team was given 3€ as a "startup capital" and they have to earn as much as possible in one week. The second one was the development of a student cafeteria. In the third project, students had to develop new products using wood as the key material (Slovenia has a large supply of wood that is being underutilized). More field work exercises and homework in customer empathy.

New components

The projects more oriented towards business-specific field exercises such as analysis of competitors, getting pro-forma invoices, negotiation, sales, and marketing in order to produce projects that were viable.

Student feedback tools

Informal interviews with students using a short survey with open-ended questions administered by the teaching team. Also, observations, individual consultations, and a formal survey where students assess the quality of the course.

Key benefits

Projects were better developed from all angles (feasibility, desirability, viability). Specific skills were trained through the exercises.

Key problems

Decreasing motivation of the students. Many indicated they were not interested in entrepreneurship as a career choice after the course. The share of projects that was continued after the course decreased year after year. The course lacked resources for more complex physical prototypes like wood products.

Lessons learned

There was a need to improve student motivation to become an entrepreneur and build on entrepreneurial training students gained from previous courses taken in students' first and second years.

Year 7

Course content & structure

In this course, students could choose a team project or an individual project. The projects lasted throughout the course. Students gave a presentation each week and were provided feedback on their progress and future goals. Teaching emphasized on product viability, feasibility and desirability. Emphasis on self-motivation, viability (business modeling) and training of specific skills such as sales.

New components

This was the first year students could choose an individual or group project, which allowed them more freedom to pursue their personal interests. Also, new prototyping rooms, with basic prototyping materials as well as wood and metalworking machinery, were introduced to provide more resources.

Student feedback tools

Informal interviews with students using a short survey with open-ended questions administered by the teaching team. Also, observations, individual consultations, and a formal survey where students assess the quality of the course.

Key benefits

As the students have had ample entrepreneurial training in previous courses, the key intent of the course was to enable the students to develop their own entrepreneurial projects and not to work on hypothetical projects, which was seen as a benefit.

Key problems

There was still lack of motivation and entrepreneurial intention to continue the projects after the course among the participating students.

Lessons learned

The faculty theorized that there was a need for a more personal career coaching approach.

Year 8

Course content & structure

The course was composed of two parts. The first consisted of individual entrepreneurial projects that were facilitated through individual consultations with the teaching team. The second part was a group project that the students had to participate in during a mandatory "Startup Weekend." The FELU Startup Weekend was modeled on an international event that is a 54-hour weekend were teams develop and design a product, business model, prototype and business pitch for a panel of judges. During the Start up Weekend, the whole faculty was transformed into a startup accelerator, with open access to prototyping workshops.

New components

The Startup Weekend served as a condensed, hands-on version of what was typically seminar work. The individual coaching was added to provide a more personal teaching approach.

Student feedback tools

Informal interviews with students using a short survey with open-ended questions administered by the teaching team. Also, observations, individual consultations, and a formal survey where students assess the quality of the course. Students prepared an individual reflection on their experience in Startup Weekend.

Key benefits

Student motivation during the Startup Weekend was very high and produced promising team results.

Key problems

The motivation for the individual project was fairly low with results below expectations of the teaching staff.

Lessons learned

The individual mentoring sessions with each student proved to be relatively ineffective. Faculty got to know the individual students but the overall quality of the individual projects was below expectations. Students were still lacking motivation even when projects were conceived by students.

redesign process repeated each academic year, including descriptions of the course content and structure, new components that were added to the course, the student feedback tools, the key benefits and key problems of the course identified by the teaching staff, and the lessons learned that were used to change the next year's course content and structure. The results presented in Table 2.1 are described in more detail later.

2.4.1 Empathize

In order to better understand the effects of the content and structure of SEB LU entrepreneurship courses as they were taught until 2006, a literature review was conducted first. An early evaluation by Drnovšek and Glas (2002) found that the number of students enrolling in the graduate level entrepreneurship program at SEB LU was rising, but there was not enough innovation in teaching methods. Students were disoriented by the incoherent use of different teaching approaches, and the classes were too focused on lectures and written seminar work. The authors suggested some changes to the program including investment in new equipment, technical training of staff, virtual communication with students, stricter student selection, and the development of specialists in different fields such as marketing and finance. The authors also suggested changes to the teaching approaches, including the addition of multidisciplinary case work, coordination of seminar work themes, updated literature, and different approaches to grading.

Another study of the undergraduate entrepreneurship course revealed that the program had no influence on the students' intentions to start their own businesses (Stritar & Drnovšek, 2006). These findings constituted a turning point for the entire SEB LU teaching team. Although the faculty knew intuitively that changes had to be implemented if future entrepreneurs were to be nurtured (as opposed to business students just being taught how to write business plans), the quantitative data from the original study was surprising.

Based on this information, the teaching staff decided to redesign the Entrepreneurial Project course for third-year undergraduate students, with the principal goal of making students more entrepreneurial, a complex and somewhat ambiguous task. Since the teaching team lacked experience in designing practice-based courses, the staff started with a systematic process for understanding the expressed and latent needs of students, which was the empathize step of the design thinking process. To develop a comprehensive curriculum and teaching methodology, the teaching team had to understand the needs and opinions of different stakeholders in entrepreneurial education including students, educators, and seasoned entrepreneurs. Therefore, the teaching team conducted a series of interviews with students, ran weekend workshops with faculty members and entrepreneurs to help map the skills and behaviors needed in successful entrepreneurship, and tested the exercises they developed to train skills and mindsets.

A main component of this process was a series of expert meetings with key stakeholders, including entrepreneurship researchers, practitioners, students, policy makers, and numerous academic guests from the arts and sciences. Structured brainstorming sessions were conducted in heterogeneous groups in order to generate ideas of what the future of entrepreneurship education should look like (Debnath et al., 2007). For example, an experienced entrepreneur commented:

> *Students had to realize that the entrepreneurial process does not start with a wish to earn money but with a clearly identified problem and a viable solution to this problem. Using observation, prototyping, and refining, you develop the actual solution and test it in real life.*

This common opinion guided the first changes to the course. At the same time as gaining insights from prospective users, the teaching team also studied and analyzed best practices and key findings from other schools that were applying similar pedagogical approaches, including Stanford University, Potsdam University, Darden Business School, University of California Berkley, and the University of Toronto, and others. Stanford University's pedagogical practices served as a model for the SEB LU entrepreneurship program. Stanford's Institute of Design, known as the "D.school", provides a curriculum on the use of design thinking for producing creative solutions to the most complex challenges, and Stanford's "REDlab" conducts research on the use of design thinking in K-12, undergraduate, and graduate educational settings.

The teaching team also engaged with so-called extreme users, which in this case was a small group of students who expressed above-average interest in entrepreneurship and starting their own venture. These students had their own entrepreneurial ideas and had a clear goal to start a company after graduation. The goal of these interviews was to understand student values and motivation in order to design pedagogical approaches that could help other students become more entrepreneurial.

The five design thinking phases of the course redesign were conducted each academic year. The empathize phase was represented by the ongoing feedback loop in this process. Each year, the teaching team gathered feedback from students on their experiences in the course and their opinions of the new components, or "prototypes", that were added to the course each year. Column D in Table 2.1 displays the various mechanisms used by the teaching team to learn from students, including observations, interviews, and formal surveys. The results of these feedback efforts were then used to identify key benefits and key problems of the course structure each year and to create new course components, or prototypes, that would be tested the following academic year. Sometimes major components of the course stayed the same, but each year there was at least one new component added.

The faculty felt that the empathize process was very beneficial to the course redesign because it provided them with information about student needs specific to SEB LU students that they would not have been able to gain from observations alone or literature on entrepreneurship education. One faculty member stated, "*With constant and, more often than not, informal feedback from students I realized that you simply cannot overlook the changing environment and specific needs of students entering the labor market*".

2.4.2 Define

The define phase of the design thinking process was characterized by the identification of key problems with the current course content and structure. The key problems identified are displayed in Column F of Table 2.1.

When the initial empathize phase was conducted, the teaching faculty found that the established mindset of the typical SEB LU business student still favored working for a large company or governmental institution. This finding was in line with research in entrepreneurship education that characterized Slovenians as risk-averse because the national culture does not encourage individual, merit-based success and independence (Dimovski & Žnidaršič, 2004; Glas, 2004). The faculty found that students felt starting their own companies would take great courage, energy, and support to go against the odds of failure to do so. Based on this information, the faculty defined the problem of the entrepreneurship program as being a lack of hands-on and applied learning. It was theorized that a practice-based course focusing on prototyping projects could help students develop stronger entrepreneurial motivation, as students would experience an applied approach to entrepreneurship learning, gain skills, and realize the potential for their business ideas. Therefore, the teaching team in the next iteration structured the course around hands-on team projects. Teams of three to five students (self-selected) worked on four different entrepreneurship projects that involved developing a new consumer product. Unlike previous courses, there was no textbook or structured content. The course method was, from the beginning, aimed at self-learning and experiential learning.

In years following, the faculty changed the content and structure of the course to test new ideas. The students were increasingly encouraged to work on meaningful projects, public presentations, the use of modern technologies and were encouraged to work intensively for shorter periods instead of sporadic work during the whole semester. Students were also increasingly given a choice on how to run their projects and who to team up with to increase relevancy and make the content relevant to real life. More and more focus was put on developing competencies rather than acquiring knowledge. As the course changed, the key problems of the course also changed. Early on, the problem was that the team projects were of low quality. Based on students' feedback, the faculty thought that this could be due to the unstructured nature of the course, which lacked focus on some key elements of business development. For example, initially the coursework was very much focused on research and development of the students' prototypes, and not enough focus was placed on helping students develop other important skills, such as identifying market opportunities, understanding customers, and sales. In later years, with more emphasis on teaching students how to understand customers, the quality of team projects improved. However, now the key problem was that students were still not particularly motivated to become entrepreneurs after the course.

The faculty found it was beneficial to specifically identify key problems at the end of the course and aim to address them by making changes to the next year's course or as one faculty member explained:

After the course was over, the teaching team had a meeting to conduct a retrospective and discuss what was "good" and what was "bad" about the course and put down a list of changes we should implement next year.

This approach is relatively uncommon in higher education where courses often stay the same year after year, even despite the feedback from students. In some cases, these new changes were very effective, such as the introduction of experienced mentors from business and academia. Mentors from smaller companies and sole proprietors were increasingly being invited to the course over mentors from larger companies. An interesting development from this was that faculty members increasingly visited various conferences where they improved their knowledge about recent developments in novel entrepreneurship pedagogies.

On the other hand, it was often a challenge to identify the cause of the key problems, like for example low student motivation and engagement. The new course components were essentially theories, and in some consecutive years, the new course components, such as the personal coaching element, were not as effective as planned.

2.4.3 Ideate

The ideate phase of the design thinking process was characterized by the teaching team identifying lessons learned based on the key benefits and key problems identified through the student feedback. The lessons learned are displayed in Column G of Table 2.1.

Much of the idea generation was on one hand inspired by what faculty experienced in the classroom as well as influenced by the faculty's interaction with other universities that were engaged in innovative approaches to entrepreneurship education, like for example Babson College and Stanford University. Early on, one of the senior faculty members interviewed for the case study attended a conference on entrepreneurship for engineering students and was introduced to the emerging program of innovative pedagogy at Stanford University. He reflected on what turned to be a pivotal event, stating:

Their [Stanford Design School's] methodology offered a different teaching approach as it was based on a trial and error prototyping approach, emphasizing interdisciplinary and innovative thinking.

This visit inspired the decision to incorporate some of the problem-solving techniques used at Stanford into the Entrepreneurial Project course at SEB LU. At that time, the teaching team had no specific literature or guidelines on how to adapt the Stanford approach and implement it into entrepreneurship pedagogy. This is how the teaching team came to use the design thinking approach as a method to redesign the course through continuous student feedback and course prototyping.

Each year, the teaching team gathered extensive quantitative and qualitative student feedback to identify key benefits and key problems in order to come up with lessons learned. These lessons learned are displayed in Column G of Table 2.1, and these were used as the basis for new course components in following years displayed in Column C of Table 2.1.

As mentioned previously, the new components were concepts that were developed through ideation, and many of the course components came from ideas and practices used at other universities. Although not a perfect process, the teaching staff as a group did the ideation, and they felt the collaborative brainstorming process helped produce new ideas that might not have been created by individual teachers or as one of the faculty members noted:

> *We had more meetings regarding this course than any other course I was involved in so far. We were discussing everything from what we did last week to what, how and why will be taught this week.*

2.4.4 Prototype

The prototype phase of the design thinking process was characterized by the teaching team implementing new course content and structure each year (displayed in Column B of Table 2.1). The key components of the course that were new are described in Column C of Table 2.1.

When changing the course content and structure, the teaching team attempted to address some or all the key problems and lessons learned from the previous year. The new course components, and the course itself, represented a prototype that was to be implemented and tested. For example, the faculty found through student feedback that the share of entrepreneurial projects that was continued after the course decreased year after year. The faculty theorized that there was a need to improve student motivation and to build on entrepreneurial training gained from students' previous courses in entrepreneurship. Therefore, the course introduced an SEB LU "Startup Weekend". The three-day event, referred to by the faculty as "condensed seminar work", was modeled on the internationally renowned Startup Weekend events. Startup Weekends are 54-hour competitions where self-selected interdisciplinary teams of aspiring entrepreneurs work on business ideas pitched by participants at the beginning of the event. Throughout the 54 hours, teams may develop a prototype or descriptive website for a product, research the market opportunity, speak with customers, and develop a business plan. They also have access to experienced business advisors throughout the event. The teamwork culminates in a business "pitch" to a panel of judges, which are often established entrepreneurs and investors, and the teams sell their product's vision, growth, and profit potential. Some Startup Weekend teams have continued working on their ideas and have launched successful startups. The SEB LU faculty felt this popular event could be a useful model for Entrepreneurial Project course. During the SEB LU Startup Weekend, the whole faculty was transformed into a startup accelerator, with open access to prototyping workshops. From the feedback at the end of the course, the faculty found student motivation during the Startup Weekend was very high and felt that the event produced promising team results. Therefore, the faculty is planning to continue the SEB LU Startup Weekend in the following years.

Several experiments were also carried out on the physical and mental environment. To create an effective environment, the teaching team increasingly offered tutoring, acquired two prototyping rooms at the campus, and created situations where students had a chance to produce physical prototypes. Students were increasingly empowered to design their own projects and to work with various stakeholders.

The benefit of prototyping new course structures and components was the ability for staff to test new ideas each year and determine which components were worth keeping or not. For example, the team project has remained a core component of the Entrepreneurial Project course throughout the eight years as it was found to very useful for practicing entrepreneurial skills. However, constant prototyping and iteration also limited the faculty's ability to identify the true benefits and problems of certain components. Some prototypes that produced poor results in an academic year may have been effective in later years with different classes of students, different faculty, extra resources, or other things. Rapid prototyping could neglect the effect of diverse teaching-learning situations and teams discussed by Booyse (2010).

2.4.5 Test

The test phase of the design thinking process encompassed the gathering of student feedback on their experience in the course (Column D of Table 2.1) and using this information to identify key benefits, key problems, and lessons learned from the course prototype.

Each year, major changes were made to the content and structure of the course, which are displayed in Column B of Table 2.1. For example, faculty members tried to identify the best combination of structured and unstructured entrepreneurial projects; therefore, one of the main aspects of the course that was changed often for testing was the individual and team projects. In the beginning, the course only included team projects, which were conducted by self-selected groups of students who worked on project ideas provided by the faulty. Two main problems with this approach were that the teams did not have an adequate mix of different skills, and the use of pre-established project ideas kept learning focused on solving problems of existing ventures, rather than helping students learn to identify new market opportunities. To address this, faculty recruited students from other disciplines to the course, in order to create well-rounded teams, and teams were also able to come up with their own project ideas. Later on, the course also included individual projects, so that students had more freedom to work on their own ideas and develop individual skills. Afterwards, the faculty chose to test a new structure, with half of the course devoted to a more traditionally designed course with seminars and written work and the other half devoted to three projects that all the students were required to participate in. At the end, the faculty tested a new structure once again and allowed students to choose either an individual or team project that lasted throughout the course.

The testing process was more difficult in some years than others. For example, some faculty noted that when 10 lectures and prototyping workshops were added to the course, the resources used exceeded the course budget. Therefore, in following years, the time and costs of testing new components became a major consideration that created boundaries for the types of course activities that could be tested.

While the testing of new course contents and structures allowed the faculty to receive student feedback on different approaches, the iteration of the course content did not necessarily achieve a "perfect" solution.

2.5 DISCUSSION

While design thinking has been recognized as an effective method for design, business, and engineering (Rauth et al., 2010), it has not been assessed systematically as a course design methodology. The case study presented extends the body of current literature in the field of course design by illustrating the methodological steps of design thinking and its benefits and challenges for course designers. From the methodological perspective, the application of the design thinking methodology to course design does not require fundamental changes in the way courses are currently designed, but it offers an additional set of approaches that might greatly improve the process. Most importantly, it enables course designers to develop in-depth understanding of how students interact with the course content and structure, which can result in course designs that are more successful in achieving their goals and are more satisfactory to all stakeholders. It also provides constant motivational triggers for further pedagogical development of the lecturer, which prevents monotony that often occurs with courses at the university level.

As presented in this case study, each step of the design thinking process for the course redesign had advantages and challenges. The empathize step was one of the most beneficial phases. Although it was quite time intensive, it allowed the teaching staff to gain in-depth information on students' needs and interests, which were used to develop theories and prototypes for the following year's course. The faculty felt that these changes improved the quality of the course as the quality of the individual and group projects improved. One of the most challenging was the test step, as in some years the implementation of new course content and structure was costly, and, in some cases, new components were not as effective as expected. However, the faculty was able to learn from these failures and identify what components to change in the following year.

The case study revealed how the application of the design thinking approach to this case of course design had many of the key aspects of design thinking described by scholars (Brown, 2008; Fraser, 2007; Nussbaum, 2004; Rauth et al., 2010). First, it was human-centered and empathetic because the teaching staff really cared about the perspectives of students, and they attempted to build a course around their needs and interests. People including students, stakeholders, and the faculty were the source of inspiration and the focus of problem solving. Second, it was iterative. Every year new course structures and components were tested, and the aim was to address key problems and lessons identified in the previous year. Third, it challenged constraints and supported creative solutions. For example, the inclusion of the SEB LU Startup Weekend as "condensed seminar" work was highly unorthodox for SEB LU, but it produced good team results.

The design thinking approach addressed some of the problems with traditional course designs discussed in the introduction. Some scholars believe traditional methods are top-down and not holistic (Van Merriënboer & Kirschner, 2013), while design thinking is focused on empathy and gathers feedback from a wide array of users and stakeholders, using that feedback to produce new and sometimes radical solutions. Also, design thinking includes prototyping and testing, which traditional methods sometimes neglect (Tripp & Bichelmeyer, 1990).

Design thinking as a course design method is especially advantageous for the development of entrepreneurship education. A major benefit of the faculty using the design thinking approach for course design was the enhancement of their own entrepreneurial mindset. The key mindsets the teaching team developed were (1) treating a course as an ever-evolving prototype, (2) empathizing with customers (students), (3) rapidly responding to different impulses from students and consequently altering the prototype, and (4) changing the role of faculty members to not just deliver content but also to offer active support to student teams. Additionally, they involved numerous stakeholders from the business community to contribute to the course design, and their involvement has increased the employability of students and provided for a wide array of applicable skills (Cox & King, 2006).

A major challenge of the design thinking process as a course design method is that even with constant prototyping and testing, the course is still not perfect. Although the extent and quality of the teaching approach have improved, the lack of entrepreneurial motivation among students remains a core problem of the entrepreneurship program. The faculty is still trying to figure out how to address this problem. This problem may be caused by factors outside of the course itself, such as Slovenian's tendency to be risk-averse and the entrepreneurial environment in and around the university (Varblane & Mets, 2010), but these aspects definitely require additional research to be better understood and addressed.

There are some limitations to this study. First, as a single case study, this faculty's experience may be very different than others using design thinking as a course design method, and these findings cannot be generalized to all situations. Motivation and discipline of university level students are different than at other levels of educational system; therefore, we would need to assess the usability of the proposed course design methodology for lower levels of the educational system. Second, course planning is part of wider curriculum planning, and we did not address the reshaping of the curriculum which would be an integral part of long-term developments in higher education (Briggs et al., 2003). Also, this study did not systematically assess the learning outcomes that came from the changes made to the course. Although the faculty felt overall that the course was improved, without measuring specific metrics of success, we are unable to conclude if the use of design thinking for the course design resulted in students gaining more knowledge, skills, or aspirations. A longitudinal study on how entrepreneurial intentions changed between students taking this course and a control group whose members did not would greatly enrich this study. This research would be important for determining the true merit of design thinking as a course design methodology. We also need to assess if tools and approaches used with traditional design thinking are directly applicable with course design and possibly develop a new tool set for course design.

2.6 CONCLUSION

Design thinking has some promise as an innovation-oriented approach to course design, particularly for entrepreneurship education. The case study shows how unlike traditional course design methods, the design thinking method is more empathetic and focused on the needs of students, is iterative, and supports creative solutions for key problems in the course. Also, faculty improved their own entrepreneurial skills. The key mindsets the teaching team developed were (1) treating a course as an ever-evolving prototype, (2) empathizing with customers (students), (3) rapidly responding to different impulses from students and consequently altering the prototype, and (4) changing the role of faculty members to not just deliver content but also to offer active support to student teams. This case study extends the body of current literature in the field of course design by illustrating the methodological steps of design thinking and its benefits and challenges for course designers.

Note: This chapter is an extended version of the paper titled *Design Thinking as a Course Design Methodology*, presented at eighth International Scientific Conference on Economic and Social Development in Zagreb, Croatia (Book of Proceedings p. 465–476).

REFERENCES

Antoncic, B., Scarlat, C., & Hvalic Erzetic, B. (2005). The quality of entrepreneurship education and the intention to continue education: Slovenia and Romania. *Managing Global Transitions*, *3*(2), 197–212.

Bae, T. J., Qian, S., Miao, C., & Fiet, J. O. (2014). The relationship between entrepreneurship education and entrepreneurial intentions: A meta-analytic review. *Entrepreneurship: Theory & Practice*, *38*(2), 217–254.

Bjorklund, T., Keipi, T., Celik, S., & Ekman, K. (2019). Learning across silos: Design factories as hubs for co-creation. *European Journal of Education*, *54*(4), 552–565.

Booyse, C. (2010). *A constructivist approach in instructional design and assessment practice* (PhD Thesis). University of South Africa, Pretoria.
Briggs, C. L., Stark, J. S., & Rowland-Poplawski, J. (2003). How do we know a "continuous planning" academic program when we see one? *The Journal of Higher Education, 74*(4), 361–385.
Brown, T. (2008). Design thinking. *Harvard Business Review, 86*(6), 84–92.
Cankar, F., Deutsch, T., Zupan, B., & Setnikar Cankar, S. (2013). Schools and promoting innovation. *Croatian Journal of Education, 15*(Sp. Ed. 2), 179–211.
Chrisman, J. J., McMullan, W., Ring, J. K., & Holt, D. T. (2012). Counseling assistance, entrepreneurship education, and new venture performance. *Journal of Entrepreneurship and Public Policy, 1*(1), 63–83.
Collins, L., Hannon, P. D., & Smith, A. (2004). Enacting entrepreneurial intent: the gaps between student needs and higher education capability. *Education+ Training, 46*(8/9), 454–463.
Cox, S., & King, D. (2006). Skill sets: An approach to embed employability in course design. *Education+ Training, 48*(4), 262–274.
Davidson, C. I., Ambrose, S. A., & Simon, H. A. (1994). *The new professor's handbook: A guide to teaching and research in engineering and science.* Bolton, MA: Anker Pub. Co.
Debnath, S. C., Tandon, S., & Pointer, L. V. (2007). Designing business school courses to promote student motivation: an application of the job characteristics model. *Journal of Management Education, 31*(6), 812–831.
Dimovski, V., & Žnidaršič, J. (2004). Entrepreneurship: An educational perspective (the case of Slovenia—compared to developed economies). *International Business & Economics Research Journal, 3*(7), 17–30.
Dogan, M., & Pelassy, D. (1984). *How to compare nations: Strategies in comparative politics.* Chatham: Chatham House.
Drnovšek, M., & Glas, M. (2002). The entrepreneurial self-efficacy of nascent entrepreneurs: the case of two economies in transition. *Journal of Enterprising Culture, 10*(2), 107–131.
Eisenhardt, K. M. (1989). Building theories from case study research. *Academy of Management Review, 14*(4), 532–550.
Eisenhardt, K. M., & Graebner, M. E. (2007). Theory building from cases: Opportunities and challenges. *Academy of Management Journal, 50*(1), 25–32.
Falvo, D. A., & Urban, M. (2007). W. Lidwell, K. Holden and J. Butler: Universal principles of design. *Educational Technology Research and Development, 55*(3), 297–300.
Fayolle, A., & Gailly, B. (2009). Assessing the impact of entrepreneurship education: A methodology and three experiments from French engineering schools *Handbook of university-wide entrepreneurship education* (pp. 203–214). Cheltenham and Northampton, MA: Elgar.
Flyvbjerg, B. (2006). Five misunderstandings about case-study research. *Qualitative Inquiry, 12*(2), 219–245.
Fraser, H. M. A. (2007). The practice of breakthrough strategies by design. *Journal of Business Strategy, 28*(4), 66–74.
Gibb, A. (2002). In pursuit of a new 'enterprise' and 'entrepreneurship' paradigm for learning: creative destruction, new values, new ways of doing things and new combinations of knowledge. *International Journal of Management Reviews, 4*(3), 233–269.
Glas, M. (2004). *The Slovenian model of SME development.* Geneva: Economic Commission for Europe.
Higgins, J. S., Maitland, G. C., Perkins, J. D., Richardson, S. M., & Piper, D. W. (1989). Identifying and solving problems in engineering design. *Studies in Higher Education, 14*(2), 169–181.

Hokanson, B., Miller, C., & Hooper, S. (2008). Role-based design: A contemporary perspective for innovation in instructional design. *TechTrends*, *52*(6), 36–46.

Hong, H. Y., & Sullivan, F. R. (2009). Towards an idea-centered, principle-based design approach to support learning as knowledge creation. *Educational Technology Research and Development*, *57*(5), 613–627.

Honig, B., & Karlsson, T. (2004). Institutional forces and the written business plan. *Journal of Management*, *3*(1), 29–48.

Huber, L. R., Sloof, R., & Van Praag, M. (2012). The effect of early entrepreneurship education: Evidence from a randomized field experiment. *European Economic Review*, *72*, 76–97.

Hytti, U., & O'Gorman, C. (2004). What is "enterprise education"? An analysis of the objectives and methods of enterprise education programmes in four European countries. *Education+ Training*, *46*(1), 11–23.

IDEO. (2014). *Design thinking for educators toolkit*. Palo Alto: IDEObooks.

Kirby, D. (2004). Entrepreneurship education: can business schools meet the challenge? *Education+ Training*, *46*(8/9), 510–519.

Kuratko, D. F. (2005). The emergence of entrepreneurship education: Development, trends, and challenges. *Entrepreneurship: Theory & Practice*, *29*(5), 577–597.

Laukkanen, M. (2000). Exploring alternative approaches in high-level entrepreneurship education: Creating micromechanisms for endogenous regional growth. *Entrepreneurship & Regional Development*, *12*(1), 25–47.

Lea, M. R. (2004). Academic literacies: A pedagogy for course design. *Studies in Higher Education*, *29*(6), 739–756.

Loughran, J. (2013). Pedagogy: Making sense of the complex relationship between teaching and learning. *Curriculum Inquiry*, *43*(1), 118–141. doi:10.1111/curi.12003

Luka, I. (2019). Design thinking in pedagogy: Frameworks and uses. *European Journal of Education*, *54*(4), 499–512.

Mahmoud-Jouini, S. B., Fixson, S. K., & Boulet, D. (2019). Making design thinking work: Adapting an innovation approach to fit a large technology-driven firm. *Research-Technology Management*, *62*(5), 50–58.

Mäkelä, M. M., & Turcan, R. V. (2006). Building grounded theory in entrepreneurship research. In H. Neergaard & J. P. Ulhři (Eds.), *Handbook of qualitative research methods in entrepreneurship* (Vol. 2, pp. 122–143). Cheltenham: Edward Elgar.

Martin, R. (2007). *The opposable mind: How successful leaders win through integrative thinking*. Boston: Harvard Business School Press.

Meyer, C. B. (2001). A case in case study methodology. *Field methods*, *13*(4), 329–352.

Mwantimwa, K., PhD. (2019). Higher institution education and entrepreneurial knowledge acquisition of graduates in Tanzania. *Current Politics and Economics of Africa*, *12*(2), 209–229.

Neck, H. M., & Greene, P. G. (2011). Entrepreneurship education: Known worlds and new frontiers. *Journal of Small Business Management*, *49*(1), 55–70.

Nussbaum, B. (2004). The power of design. *Business Week*, *17*(2), 86–94.

O'Brien, E., & Hamburg, I. (2019). A critical review of learning approaches for entrepreneurship education in a contemporary society. *European Journal of Education*, *54*(4), 525–537.

Oosterbeek, H., van Praag, M., & Ijsselstein, A. (2010). The impact of entrepreneurship education on entrepreneurship skills and motivation. *European Economic Review*, *54*(3), 442–454.

Piperopoulos, P. (2012). Could higher education programmes, culture and structure stifle the entrepreneurial intentions of students? *Journal of Small Business and Enterprise Development*, *19*(3), 461–483.

Platt, J. (1992). Case study in American methodological thought. *Current Sociology, 40*(1), 17–48.
Posner, G. J., & Rudnitsky, A. N. (1994). *Course design: a guide to curriculum development for teachers.* White Plains, NY: Longman.
Pretorius, M. (2008). Assessment of entrepreneurship education: A pilot study. *Southern African Journal of Entrepreneurship and Small Business Management, 1*(1), 1–20.
Rauth, I., Köppen, E., Jobst, B., & Meinel, C. (2010). *Design thinking: an educational model towards creative confidence.* Paper presented at the Proceedings of the 1st International Conference on Design Creativity ICDC 2010, Kobe.
Rutherford, S. (2020). Engaging students in curriculum development through design thinking: A course design case. *International Journal of Designs for Learning, 11*(3), 107–125.
Schweitzer, J., Groeger, L., & Sobel, L. (2016). The design thinking mindset: An assessment of what we know and what we see in practice. *Journal of Design, Business, & Society, 2*(1), 71–93.
Shawer, S. F. (2010). Classroom-level curriculum development: EFL teachers as curriculum-developers, curriculum-makers and curriculum-transmitters. *Teaching and Teacher Education, 26*(2), 173–184.
Stanford University, D. S. (2007). *Hasso planter: Institute of design at Stanford.* Retrieved from www.stanford.edu/group/dschool/
Stenhouse, L. (1988). Case study method. In J. P. Keeves (Ed.), *Educational research, methodology, and measurement: An international handbook.* Oxford: Pergamon Press.
Strauss, A., & Corbin, J. M. (1997). *Grounded theory in practice.* Thousand Oaks, CA: Sage.
Stritar, R., & Drnovšek, M. (2006). *Dejavniki oblikovanja podjetniških namenov študentov: vpliv timskega dela: magistrsko delo.* Ljubljana: Univerza v Ljubljani, Ekonomska Fakulteta.
Timmons, J. A., & Spinelli, S. (1994). *New venture creation: entrepreneurship for the 21st century* (Vol. 4). Boston: McGraw-Hill/Irwin.
Tripp, S., & Bichelmeyer, B. (1990). Rapid prototyping: An alternative instructional design strategy. *Educational Technology Research and Development, 38*(1), 31–44.
Tyler, R. W. (1950). *Basic principles of curriculum and instruction.* Chicago: University of Chicago Press.
Van Merriënboer, J. J., & Kirschner, P. A. (2013). *Ten steps to complex learning: a systematic approach to four-component instructional design.* New York: Routledge.
Varblane, U., & Mets, T. (2010). Entrepreneurship education in the higher education institutions (HEIs) of post-communist European countries. *Journal of Enterprising Communities, 4*(3), 204–219.
Visscher-Voerman, I., & Gustafson, K. L. (2004). Paradigms in the theory and practice of education and training design. *Educational Technology Research and Development, 52*(2), 69–89.
Walvoord, B. E., & Anderson, V. J. (2009). *Effective grading: A tool for learning and assessment in college.* San Francisco: Jossey-Bass Inc Pub.
Ward, A., Runcie, E., & Morris, L. (2009). Embedding innovation: Design thinking for small enterprises. *The Journal of Business Strategy, 30*(2/3), 78–84.
Whetten, D. A. (2007). Principles of effective course design: What I wish I had known about learning-centered teaching 30 years ago. *Journal of Management Education, 31*(3), 339.
Yin, R. K. (2003). *Case study research: Design and methods.* London: Sage Publications.
Zheng, D. (2018). Design thinking is ambidextrous. *Management Decision, 56*(4), 736–756.

3 Design Thinking—Innovation Approaches and Challenges Toward Product Design

Ranjan Kumar and Kaushik Kumar
Birla Institute of Technology, Mesra, Ranchi, India

CONTENTS

3.1 INTRODUCTION

Over the last few decades, the value of design has been understood, and the utilization of design thinking in various walks of today's technological world has

DOI: 10.1201/9781003189923-5

been regarded as an efficient tool for developing sustainable design [1]. Thus, design thinking is being regarded as a "design-based problem-solving approach for humans", which is being treated as an efficient approach for developing innovative design-based environments [2]. In recent decades, the utilization of such "design-thinking"-based design tools has been increasingly used by many of the multinational companies that are now dominating their technological sectors. According to studies, only a small percentage of managers are content with their companies' innovation initiatives, a view that reflects the fact that only about 10% of new products and services succeed [3]. Such circumstances are demanding newly fresh ways of thinking of innovations. In this regard, various schools, consultancies, and design companies have taken initiatives and have provided alternative ways of design thinking instead of running under traditional thought processes for innovations. The said transformations have been gained and responded through books, courses, and various projects [4–5].

In today's scenario, it is a fact that design thinking approaches and strategies possess various business adoptions, but the full-fledged results can only be achieved if we have the proper understanding of design thinking, their working principles, their innovative capabilities, and the circumstances under which they can be utilized. The cognizance of such understandings is essentially required to develop proper design-thinking-based approaches. Learning is a cumulative and gradual process, and, as a result, learning efficiency is significantly influenced by learning frameworks and the depth of prior knowledge [6]. Though there are a growing number of articles are being produced on design thinking, they are mostly anecdotal or prescriptive in nature and also lacking in research-based insights [7]. Various reviews also highlight the limitations involved in design thinking [8–9]. The continuous research has also highlighted a wide application arena of design thinking approaches and their underlying principles that possess many organizational capabilities [10–13]. Further, many studies have highlighted and have come across some resourceful tools such as alliance management, internal rewarding, customer participation, and some common techniques, which are pretty much helpful in developing a design thinking approach for a sustainable product design [14–17]. Further, "Innovation management is a complex phenomenon with strong interdisciplinary characteristics, and its practice permeates many of an organization's focuses and functional activities" [18–19]. The works under design thinking approaches are mostly being treated as being conceptual and exploratory, which require a better understanding of the previous four said aspects on basics of design thinking [8].

Several studies have enlightened the product innovation strategies toward the new product success (NPS) factors. However, the innovations are risky and expensive in terms of product innovations [20], but the new product innovations (NPIs) are essential due to the "high failure rate and the increasing number of new product development (NPD)". Thus, product innovation is being increasingly recognized as being a critical component of a company's long-term success [21]. Product design broadly involves a multidisciplinary approach and possesses various aspects, but at the same time, it possesses a common goal of developing a new product under industrial and engineering disciplines [22]. Various studies have been carried out in the domain of design-thinking-based product design and product innovation approach and process. Also, many literature-based recent developments and advancements in utilizing the design thinking and product innovation concepts as the hands-on tools and techniques have been depicted graphically in Figures 3.1 and 3.2.

Similarly, various contributions in conceptual design thinking as a "dynamic multi-dimensional construct, confirmation of design thinking's positive performance influence, and determination of its robust effects across varying turbulence levels" [1] are useful in providing an integrative view over design thinking product innovation (DTPI). The present work is a noble attempt that will enlighten the reader with the design thinking approaches, their basic principles along with

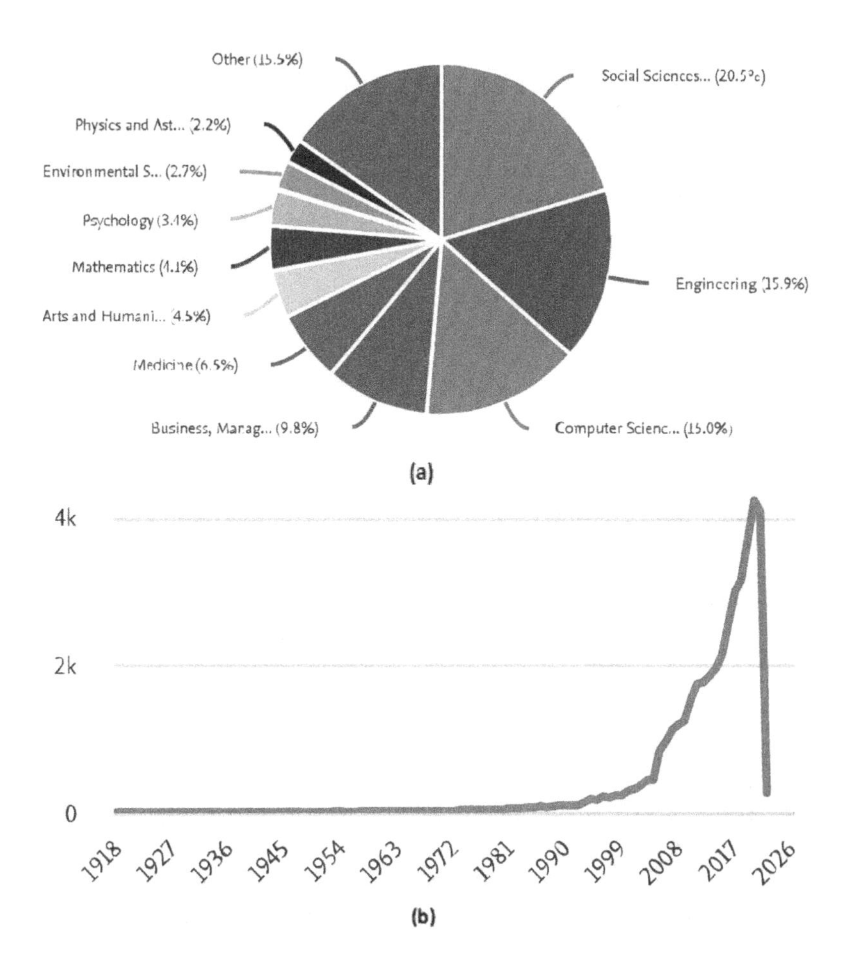

FIGURE 3.1 Various documents published in the domain of design thinking: (a) subject area wise publication and (b) year-wise documents publication [23].

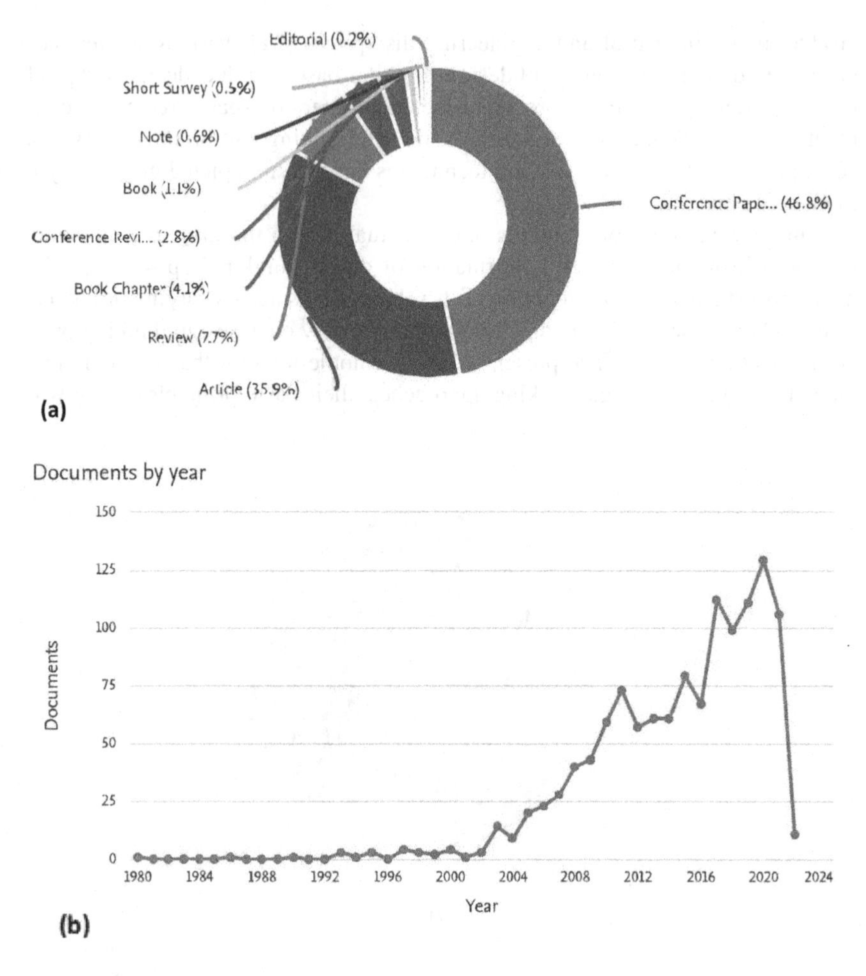

FIGURE 3.2 Various documents published in the domain of design-thinking-based product innovation: (a) types of documents published and (b) year-wise documents' publication [23].

the associated challenges during implementation. The current work is organized as follows: the chapter starts with the brief introduction of design thinking, and subsequently the basic background of design thinking has been discussed. Moving ahead to the next section, various product design approaches have been summarized briefly. Subsequently, in the next section, the DT applications in different domains have been briefly discussed; further, the associated challenges during the implementation of design thinking approaches have been summarized, and finally the chapter is concluded with highlighting the possibilities of future research trends in the future scope section.

3.2 BASICS OF DESIGN THINKING

3.2.1 Definition and Background

Design thinking has become a trend, and it is being used for innovation by everyone from students to designers, tiny studios to big corporations. Kelley and Brown introduced the notion of design thinking to business in the 1990s. Through Stanford Design Centre, the former strongly pushed its application and practice. Design thinking, as a popular method, improves an organization's ability to innovate and solve more complex business problems by incorporating ideas, interests, and values into the process [24–27]. Therefore, design thinking can be viewed as "logic, principles, practices, tools, discourse, philosophy, and mental model", helpful in getting problem-solving skills for various industrial complex design processes [28–29]. According to Zheng [13], design thinking (DT) is a "methodology which is driven by design philosophy", whereas Martin et al. [30] simply define the DT as a managerial problem-solving technique. As DT exhibits problem-driven traits, some researchers provide more emphasis on defining the problem, but some others focus more on solving the problems [31–32]. Further, according to Carlgren et al. [33], the design is all about thinking of problems, their associated challenges, and the solutions. The reframing of the problems is quite important because sometimes it is been observed that the problem is not clearly identified or is tough to define (called wicked problems). Hence, to deal with such a scenario, the problems must be reframed or redefined [34–35]. Referring to Chen et al. [36], design thinking is nothing but the "implementation of design philosophy into design processes and output".

Despite the fact that design thinking is multidimensional, it has escaped a comprehensive yet concise conception, which would obstruct empirical study and scholarly knowledge [10, 37]. The proper definition of design thinking is still not found because its conceptual utilization and understandings vary from people to people. Hence, for having a useful, consistent, and logical understanding of DT, we first need to define the DT. This is done on the basis of commonalities achieved from different definitions [1] in the tabulated form in Figure 3.3. Design thinking is frequently described as a method that depends on designers' approaches to solving very human challenges [4, 8–9, 24]. In this regard, the remarks on design thinking given by IDEO's CEO Tim Brown [4] seems to be prominent: "it is a discipline that uses the designer's sensibility and methods to match people's needs with what is technologically feasible and what a viable business strategy can convert into customer value and market opportunity". Hence, out of various available definitions from various perspectives, we propose a precise definition of design thinking as a tool for developing new products by resolving essential related problems. Design thinking comprises two things: Design, which includes various design-related activities, and thinking, which includes various perspectives of analyzing that design. Thinking is an essential part, which provides a pathway of looking at the designs from various perspectives and giving various philosophies and thoughts about the proposed design [38]. During various thought

Authors (Year)	Method	Research Topic	Definition of DT	Composition of DT	Consequence of DT
Beverland et al. (2015) (10)	Qualitative cases	DT as aid to managing tensions of brand ambidexterity	"Inherent logic and practices of designers" (p.590)	"Creative and strategic process characterized by...abductive reasoning, iterative thinking and experimentation, holistic perspective and human centeredness" (p.593)	Management of brand Consistency and relevance
Carlgren et al. (2016) (64)	Qualitative cases	Barriers to implementing DT	"Multidisciplinary, human-centered innovation an approach inspired by designers" (p.344)	"Set of principles/mindset, practices, and techniques...user focus, problem framing, visualization, experimentation, and diversity" (p.346)	DT Implementation
Chen et al (2018) (33)	Student experiment	Teaching design thinking via design techniques to produce marketing outcomes	"Implementation of design philosophy into design processes and outputs" (p.176)	User centered design technique	Perceived usefulness, originality, and value of a student idea using a specific design method
Cooperrider (2010) (11)	Conceptual	How corporate strategy and design thinking can come together to solve the most pressing sustainability issues e.g., global poverty	None	None	None
Elsbach and Stigliani (2018)(12)	Conceptual	how design thinking relates to organizational culture" (p.2278)	"an approach to problem solving that uses tools by designers of commercial products, processes, and environments" (p.2274)	Four phases; tools in each e.g. mapping, brainstorming, rapid prototyping; and thinking styles e.g. convergent vs. divergent	Organizational culture
Liedtka (2015) (7)	Conceptual	"review of assumptions, principles, and key process tools associated with DT" (p.925)	"Bringing designers' principles, approaches, methods and tools to problem solving" (p.926)	Various models of DT as mostly 3 stages and some common tools e.g. collaborative sensemaking and assumption surfacing	Reduced cognitive biases e.g., undervaluing novel ideas and overlooking disconfirming data
Micheli et al. (2019) (9)	Conceptual	Review of literature to identify principle attributes and tools/methods and future research	Notes three most commonly used definitions	Nine primary attributes e.g. iteration and eight tools/methods e.g., visualization	None
Luotola et al. (2017) (68)	Qualitative cases	Design thinking as a means of arriving at a sales solution	"discipline that uses the designer's sensibility, knowledge, and methodological understanding to match customer's needs to what is technically and economically feasible" (p.62)	Problem identification, designing a sales solution, reaching final sales or certainty	Sales certainty
Seidel and Fixson (2013) (27)	Qualitative cases	"How novice multidisciplinary teams use design methods to successfully develop novel concepts?" (p.20)	None	Need finding, brainstorming, and prototyping	Concept novelty
Zheng (2018) (13)	Conceptual	Relation of DT through ambidextrous learning ultimately to ambidextrous innovation	"Methodology driven by design philosophy" (p.738)	Traits of problem-driven, stakeholder focus, holistic and systematic perspective, experimentation, visualization, abductive reasoning	Balance of radical to incremental Innovation in a firm portfolio

FIGURE 3.3 Recent studies on design thinking published in business literature [1].

processes and different perspectives on design, the mind plays a vital role; hence, as a result, we believe that design thinking is fundamentally made up of both mindsets and cognitive actions.

3.2.2 Psyche of Design Thinking

For creative and strategic cognitive actions against the proposed designs, it is important to have a critical mindset in order to analyze the design. For such a strategic process, the literature highlights three hallmarks: human-centeredness, abductive reasoning, and iterative thinking and experimentation [10]. Mindset is an integral and distinct part of design thinking.

Design thinking exhibits its human-centered approach for examining and understanding the customer's perception and their facing issues and accordingly helps in providing the appropriate solutions. Before examining prospective solutions, DT evaluates not just the function and use of a product or service, but also the associated emotions or meanings. So, DT can be regarded as a '*user-centered approach*', which is distinct from customer orientation [24, 39–40] including a "full range of emotional, embodied, and material events" [10]. This mindset helps in creating new product innovations. As a result, the customer preferences are translated into memorable and meaningful experiences that become an integral part of the innovation itself, thanks to the human-centered or user-centered approach.

Abductive reasoning is a kind of mindset that is different from deductive reasoning and inductive reasoning, involving considerable possibilities and challenges. Therefore, it is essential for achieving new insights and cognizance [40] as "designers focus on future solutions where they perceive reality and culture as something pliable—their attitude towards workable solutions" that is "assertion-based rather than evidence-based" [41]. Other innovative methodologies, such as deductive or inductive reasoning, arrive at answers by carefully constructing premises, usually to analyze and defend a project or design decision [42].

As a learning process, Beckman and Berry's [43] model of innovations states that design thinking is a dynamic and iterative way of understanding the challenges and exploring solutions accordingly. Trial-and-error learning through iterative forms, prototyping, and trials that test a range of possible solutions with end-users and other project stakeholders are hallmarks of design thinking. The stages of innovations are also associated with failing during the learning process, and this "learning by failing" can be regarded to be another mindset in the design thinking approach. Seeing failure is a crucial part of the learning process. Daring to fail in order to learn allows you to arrive at more effective solutions faster. If one learns by failing, one understands that risk-taking—and, ironically, mistakes—can lead to unanticipated answers rather than the "careful stepwise reduction of errors" [1, 44].

3.2.3 Involved Actions

Literature surveys have identified various activities involved in design thinking. These DT-actions provide many research questions related to the identification of new product-based innovations and inventions. These actions are collectively the progressive steps toward new product development. Various researchers have proposed various DT-actions. For instance, Interaction Design Foundation [45] has proposed five activities as "empathize, define, ideate, prototype, and test". Further, Liedtka [46] has proposed some research questions such as "what is? what if? what wows? and what works?". Three important fundamental actions like "understanding the innovation need, generating potential answers, and testing and refining to arrive at the answer" have been proposed by some researchers for developing an innovative design thinking approach [2, 28]. Hence, summarizing the all-aforementioned fundamentals of DT, we can propose the basic four strategical approaches of design thinking principles as Empathy, Defining Problems, Ideation, and Solutions as depicted in Figure 3.4. Design thinking is a multidisciplinary field that is supported by the knowledge of multiple domains for the sustainable product design/development and product innovation toward the improvement of the experience of the users. The various supported domains can be viewed in Figure 3.5.

3.2.4 Working Principle

Design thinking helps people to enhance their capabilities to innovate by adopting and practicing certain mindsets. The associated gaps in DT-based approaches have been summarized in the previous section. Further, stepping ahead, the working principles of DT plays a vital role in the successful implementation of DT. The basic four strategical approaches of DT produce a drastic change during DT-based new product design. Hence, outlining the relationships among these four approaches is essential.

The predefined mindset hypothesis is influenced by each cognitive thinking and cognitive action, and these behave as a significant factor in determining

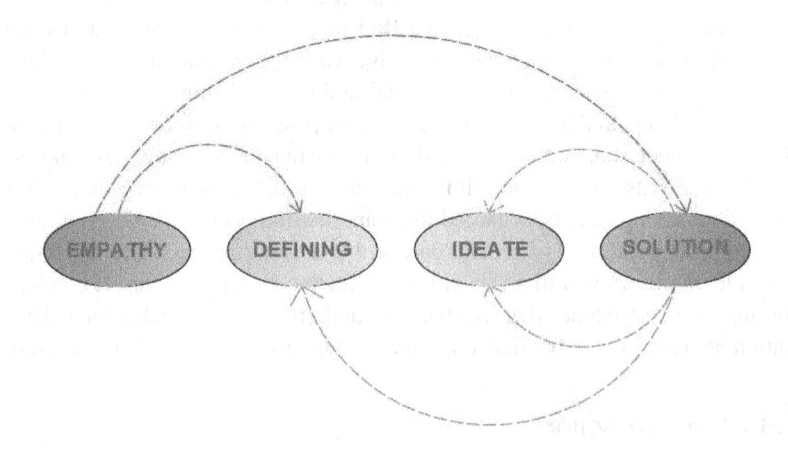

FIGURE 3.4 The schematic diagram showing the working principle of design thinking (DT).

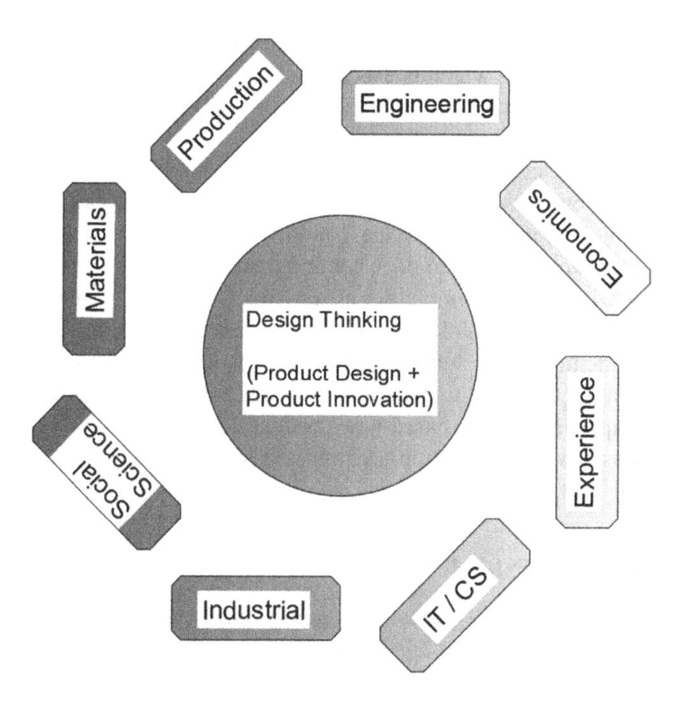

FIGURE 3.5 The schematic diagram showing the requirement of knowledge of various domains for DT.

the new innovative product designs [47], "while the tendency to engage in and enjoy thinking is an antecedent of individual innovative behaviors" [48]. Among four fundamental DT strategical approaches, Empathy is one of the most significant principles in DT-based-NPD. Empathy talks all about understanding the customer's point of view toward the day-by-day requirements of one's target audience. "Designing with empathy incorporates doing the majority of that and going an additional step. It requires really envisioning the experience of work, learning, and critical thinking from the audient's point of view". Similarly, the second working principle talks about defining the problem statement. This stage is all about taking great care in the customer's context. At this stage, we need to examine our own perceptions and integrate them in identifying and characterizing the central portion of issues. Further, the assembly of plans, functions, and other components is considered carefully to execute against the identified design issues. Stepping ahead, in the third stage of DT working principles, the ideation is done by doing various brainstorming exercises and activities in order to get the new creative solutions toward the identified design problems. In this stage, each and every thought is treated as a fantastic thought, and "nobody's thoughts are rejected". Discussions are held, and the idea session continues till getting the best possible idea for resolving the issues. Further, at last, the final DT principle is to provide the necessary solutions on the basis of the research done in all aforementioned previous three stages. The research done in previous all three stages is reflected back in the solution stage which is executed in terms of prototyping and experimentations as per the outcomes obtained from the previous stages of DT principles. The prototype could be a sketch, a model (virtual or real), or any creative technique utilized to convey the concepts of solutions, but it could have limitations in terms of budget. Similarly, experimentations involved various testing as per the data obtained in previous stages, and the purpose of various testing is to summarize what works and what does not and then to iterate the final results.

Thus, as a whole, after completing all possible four stages, one could reach the new innovative design thinking solutions toward the problems. Finally, when we start implementing the solutions, it may be proved to be very much useful to empathize with the target customers or, sometimes, situations. Such a systematic and strategic approach of DT-based new product development and new product innovations has a wide application arena in today's product development fields [49–51].

3.3 PRODUCT DESIGN

The term product design (PD) is a very "generic term", and often the people get confused because the term "*product design*" proposes a significant meaning and implied relationships between "*engineering design*" and "*industrial design*". In various cases, PD is all about related to the Engineering Design (ED), whereas in many cases, PD talks about a particular subject area of Industrial Design (ID) [52–53].

Various researchers have described the idea of PD in different ways. Roozenburg [54] described product design as framing and laying down the ideas or plans necessary to manufacture a product. Further, Horvăth [55] demonstrated the relationship among product design, industrial design, and engineering design. The design thinking contains some part of engineering design perspective and industrial design perspective that can be seen in Figure 3.6. However, the domains, ED and ID, are not only two different areas along with their own fundamentals and distinct characteristics, but also they involved some part in the product design process. Hence, neither ED nor ID can completely describe the PD processes.

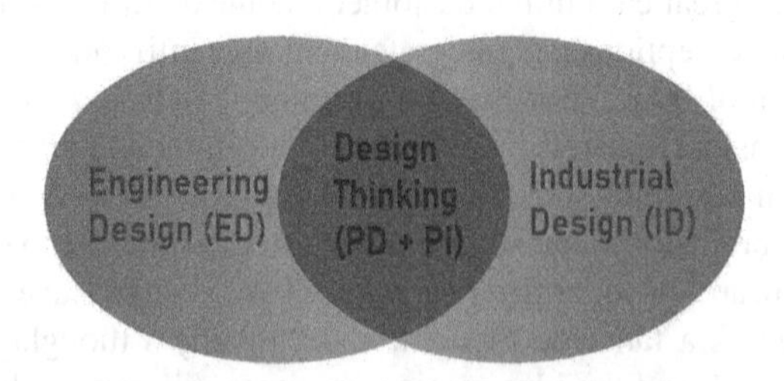

FIGURE 3.6 The schematic drawing showing the overlapping of the two domains ED and ID with design thinking (PD and PI).

The PD is not an isolated process, but it is a part of a product development process that comprises several disciplines to design and develop a complete product. During a product development process, several designers such as "industrial designer, the mechanical designer, the electronics designer, the purchasing specialist, the manufacturing engineer, and the marketing professional" perform their own specified assignments. From the professionals' perspective, the product design is an "electro-mechanical product of modest complexity which covers a significant portion of the consumer product market these days" [54, 56].

3.3.1 Product Design Approaches

As briefly discussed before, PD contains two main design domains, viz. ED and ID. PD possesses various design approaches, and it is essential to understand these approaches in terms of product design and design thinking. Reviewing from a different perspective, both the domains containing the design process are necessary. Design is all about providing the necessary solutions to human problems. Every design starts with various perspectives, and through various stages of design processes, a final solution is obtained [22]. However, depending on the design discipline, different design techniques or processes are adopted.

3.3.2 Two Approaches of Product Design

The two design perspectives, i.e., the Engineering Design (ED) and Industrial Design (ID), are involved in almost all the design processes. Most of the design models are the results of the development of "layout and form design" from the engineering perspective. However, many people suggest that the layout should be designed first before the form design, which implies that the inside design should be designed first, and, on the basis of this inside design, the outside design should be processed. But the Industrial Design perspective is completely reversed in this manner and suggests that the outside design should be processed in advance, and, on the basis of this outside design, the engineer's job is to develop an appropriate inside design. Thus, inside design can be regarded as a "design activity for defining and arranging functional components on the basis of functional structure". This is one of the types of design activity similar to layout design, which is done by engineering designers who focus immensely on product functionality. Similarly, the other design type is developing an outside design that mainly focuses on the shape, size, form, and color for their external visualization. This helps in developing human interactions and aims at providing the customers with a tremendous user experience. Hence, this can be regarded as the second type of design activity with the help of human interactions. Therefore, on the basis of the earlier discussions, the two designs can be distinctively adopted as "*Inside-Out Design Approach*" (IODA) and "*Outside-In Design Approach*" (OIDA) [22].

Further, the engineering design perspective possesses very much similarity with the industrial design perspective, but the engineering designs are developed in a very much detailed manner with keeping in mind lots of important observations and considering the various inevitable circumstances. Phal and Beitz [57] proposed an engineering design model for the design and development of any product that contains seven steps of design which are divided into four phases viz., Clarification of the Task, Concept Design, Embodiment Design, and Detail Design. This proposed model produces layout and forms development process considerations. While looking for a viable design solution, challenges are recognized, and, accordingly, the functions are structured during the conceptual design phase. The functional elements are arranged after the concept has been developed. This stage is termed as "layout design stage", where the engineers are employed to produce sustainable designs using systematic or mechanical methods for the development of best-suited layout design. As a result, the layout becomes a critical aspect in deciding the product's final form. Layout and form are finalized at the embodiment design phase.

Further, Roosenburg [58] proposed a similar type of engineering design model for the sustainable development of any product as seen in Figure 3.7. The proposed model talks about six stages of any product design and development process in which there are stages from problem identification stage to solution, and these various stages have to undergo various design proposals and applied conditions. Finally, after the design evaluation stage, the final problem-free sustainable collaborative [59] product is obtained for providing the better user experience.

3.3.3 Inside-Out Design Approach

The term "inside-out" was first used by Dreyfuss and Lorenz [60] to describe the work done by designers. The main focus of referring to this word is to design the exterior of a product from "inside-out". This IODA is a traditionally used design approach where the already developed inside functional components are provided to designers, and accordingly the outside design forms are developed [22].

3.3.4 Outside-In Design Approach

Further, the outside-in design approach solely depends upon the user's experience and nowadays contains many activities. This includes not only the outside design form but also undergoing various other design aspects based on the "user facing design" (UFD) experiences. Such UFD experiences even contain emotional factors which are considered to be one of the critical things during the OIDA approach. For enhancing the user experience related to a particular product, the designers have to focus on various significant functions related to the user experience [22, 60]. This OIDA is a rare design approach in practical cases. Because functional parts usually need a certain space to be placed and a specific style of interaction among them to perform a given purpose, interior design is barely finished using this approach. So far, we've mostly seen this kind of production in nonpractical fields like design competitions. For the objectives

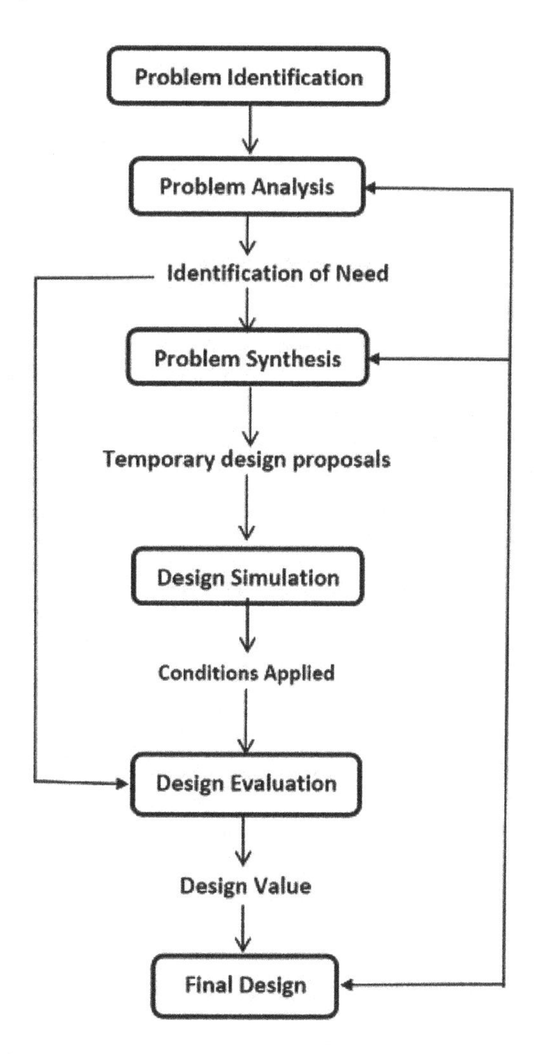

FIGURE 3.7 Engineering design process model proposed by Roosenburg [58].

of conceptual design experiments and design competitions, industrial designers have publicized pure conceptual items with very minimal and sometimes even fictitious ways of operating. This type of product frequently brings completely new concepts and, if achieved, would be extremely revolutionary. However, realizing these concepts is frequently difficult and necessitates the creation of new processes or technologies.

This appears to be a growing trend among producers nowadays. They begin product design by completing the exterior design before moving on to the interior structure. This scenario appears to be aided by technology, as many goods pertaining to outside design begin with electronic components. Small electronic components' functional elements, on the other hand, have fewer arrangement constraints than mechanical components. This may be achievable for mechanical elements and products, depending on the level of technological advancement.

3.3.5 Product Design Based on Product Type

The inside-out technique makes it easier to develop products with mechanical components. This could be due to the fact that a mechanical component can only occupy a certain specific space to make the product functional. In today's scenario, the market leaders are mostly taking an outside-in approach. Design approaches can be employed differently depending on the product type to be developed in this case. As a result, we need to categorize different products so that we may dig deeper to explore the discussed two approaches of design.

3.4 INDUSTRIAL APPLICATIONS OF DESIGN THINKING

Design thinking advocates some strategic planning with user-centered approaches toward product innovation and design with the help of the designer's thought process and their works and helps in solving human problems. The engineering design and industrial design process are the two main perspectives of any design process that allows us to design and develop a new product or produce product-based innovations for doing things easily. However, the tremendous growth in DT has provided a very much sophisticated approach in developing new products and problem-solving methodology which are being nowadays are applied in various domains.

3.4.1 Design Thinking Application in Aviation Industry

When the aerospace industry's innovation stalled, Airbus' innovation manager, Bastian Schaefer, thought it was time to get solutions for old challenges with fresh eyes. His team contacted Autodesk, and other design and engineering companies, hoping to direct the aerospace company in new ways.

The two companies collaborated to figure out the real issues and passengers' need, and they focused on making the flight journey as much comfortable as possible. However, the term "*comfortable*" is different for different people of different mindsets. "Our task is to address all of these elements in a flexible, dynamic way", Schaefer explains. The teams began finding chances to improve the plane design while enhancing the flight experience by focusing on the human-centered approach. Hence, design thinking helped them in discovering the genuine problem and their feasible solutions by using the latest technologies immensely [61].

3.4.2 Design Thinking Application in Pharma Industry

Regarding the context of healthcare system, design thinking is expected to provide a simple and meaningful patient experience. Pharma industries need to be responsive toward the patients' care, patient's behavior, and personalities. Keeping the user experience at the forefront, the final product or service must convey emotionally, which necessitates a cultural shift in how many pharma professionals think. Brown quotes a comment from the IDEO website for rigid rational thinkers who rely only on quantitative data in making decisions: "Nobody wants to run an organization on feeling, intuition, and inspiration, but an overreliance on the logical and analytical can be just as problematic".

Design thinking becomes difficult in implementation because the organizations aren't usually built with the goal of considering a patient's journey. According to Hartmann, "the customer experience is often just a by-product of the service provider's organizational structure and operations" [62].

3.4.3 Design Thinking Application in Education Sector

Design thinking is a way of thinking and develops a design mindset. It helps in infusing self-confidence in people for achieving a better desired future, as well as provides a method for taking action when confronted with a challenging task. In education, that type of optimism is desperately required.

Every day, from teacher feedback systems to daily routines, classrooms and schools around the world face design issues. The obstacles that educators face, no matter where they fall on the scale, are genuine, complex, and varied. As a result, they necessitate fresh viewpoints, tools, and techniques. One of them is design thinking. Hence, pharma industries may think of using the design thinking techniques for developing a straightforward experience of their patients without bothering or overwhelming them with lots of medical information, procedures, or jargon. According to Petroff, a human-centered approach to pharma can successfully capture issues. "I'd employ design thinking to help interdisciplinary teams better outline the problem space". He suggests, "It's a terrific technique for stakeholder alignment. [63]".

3.5 INDUSTRIAL CHALLENGES OF DESIGN THINKING

As discussed in previous sections, design thinking is a human-centered problem-solving methodology that provides very sophisticated tools and techniques to solve human problems with strategic planning and management. However, the domain of DT has grown drastically over the last few decades. However, the successful implementation of the DT process and the PI process is very effective and user-friendly in providing a better user experience. Despite being increasingly promoted as being an innovative approach, some indications from various firms suggest some observed barriers and challenges during their successful implementations.

In simple words, the DT could be understood as a "*way to solve various challenges*" [4, 60, 64–65]. However, very few works have been devoted to this field of analyzing industrial challenges during the implementation of DT in organizational settings. The notable argument was first produced by Lindberg et al. [65], who noted that implementing the DT and PI approaches is difficult because they necessitate a conceptualization process to address early inconsistencies with the organizational context and previous work [66–67]. DT is still under investigation for the successful implementation of conceptual, resourceful, and sustainable solutions. According to Walters [63], the ambiguous nature of DT is often in conflict with, and different from, dominant organizational cultures and processes. However, Carlgren et al. [68] "argued that the design thinking challenges existing design functions, as the concept could be interpreted as *everyone can do design*". The way of understanding the DT challenges during the implementation of DT can be illuminated in discussing several barriers associated with this design thinking approach. These may be classified as [8, 69] given in the subsequent sections.

3.5.1 Hurdles to Innovation

Several contributions from several authors have made the literature flooded with the existence of innovation hurdles in giant industrial firms. There exist basically two types of barriers, imposed on innovation as internal and external barriers. The internal hurdles include limiting mindsets, a lack of skills, insufficient resources, and unsupportive systems. In the same line, in another review, focusing on internal barriers only, Assink [69] proposed a conceptual framework of inhibitors to disruptive innovation, based on different hurdles as discussed later in the chapter. Outside the all-internal barriers of an organization, there exist some external barriers, which are external to the organization.

3.5.2 Hurdles to Adoption

According to Öberg and Trifilova [70] and Story et al. [71–72], the traditional mindset and routines are a major hurdle in terms of the new and innovative product development process. It is necessary for a company to break from the old and traditional design process and mindset so that a new radical idea and innovation can be born in the environment. The term "adoption" is based on rigid, existing, and dominant designs that emphasize one's area of expertise that one does not want to leave willingly. "Assink [69] also stresses the lack of appropriate processes to handle disruptive innovation along with current development". Most of the major companies suffer from a scarcity of resources for innovation, and their procedures are usually set up for incremental growth [65].

3.5.3 Hurdles to Mindset

The mindset hurdles include both individuals as well as organizational mindsets. Assink [69] focuses on "the inability to unlearn". Individuals who work in the field of innovation must have an interpretative rather than an analytical attitude, and they must also be tolerant of uncertainty. Such values of individuals may clash with the organizational mindset, and the fact speaks that the values are difficult to change.

Further, the challenges occurring in the implementations of design thinking may also be identified in the following seven types [69], which are directly interconnected between the utilization of the design thinking approach and mainstream operations, as well as the obtained output result. These changes can be given as:

- Mismatch with an existing structure.
- Implementation difficulties faced from new innovative ideas or concepts.
- Lacking the collaborative process in the workspace.
- Difficulty to prove the value of design thinking.
- Clash on individual and organizational mindsets.
- Difficulties in acquiring a new set of skills for NPD and NPI.
- Existence of various communication barriers.

3.6 CONCLUSION

In a traditional company, design is considered to be a manager's domain, who can identify a problem as a problem with several answers to choose from, or as a problem for which a solution needs to be developed. Design, on the other hand, should not be limited to managers only, but it should permeate a complete organization and must be available to any individual or group with decision-making authority.

DT is a human-problem-solving methodology that allows decision-makers to come up with innovative and distinctive solutions to challenges, and it works best when combined with the design mindset. A cognitive thought process less depends on heuristics and is regarded as the most difficult part of addressing a problem as it involves developing the best answer rather than choosing from a list of options. In conclusion, in the present work, the authors have tried identifying the design questions and have contributed by proposing a new conceptual perspective on design thinking based on product design and product innovation. During the discussion, the basic understanding and innovative design thinking approach through conceptual and exploratory design approaches have been summarized. Further, the discussion takes us to the application purview of NPD and NPI using the design thinking approach that summarizes the various domains that are interactively involved in the utilization of design thinking as a problem-solving tool. Further, the current work mainly focuses on industrial product design and product innovation using a design thinking approach. In this context, the two design approaches viz. IODA and OIDA have been discussed along with the hurdles and challenges associated with design thinking approaches.

Hence, the multidisciplinary approaches of design thinking for product design and development can be estimated from the data provided in the Figure 3.8.

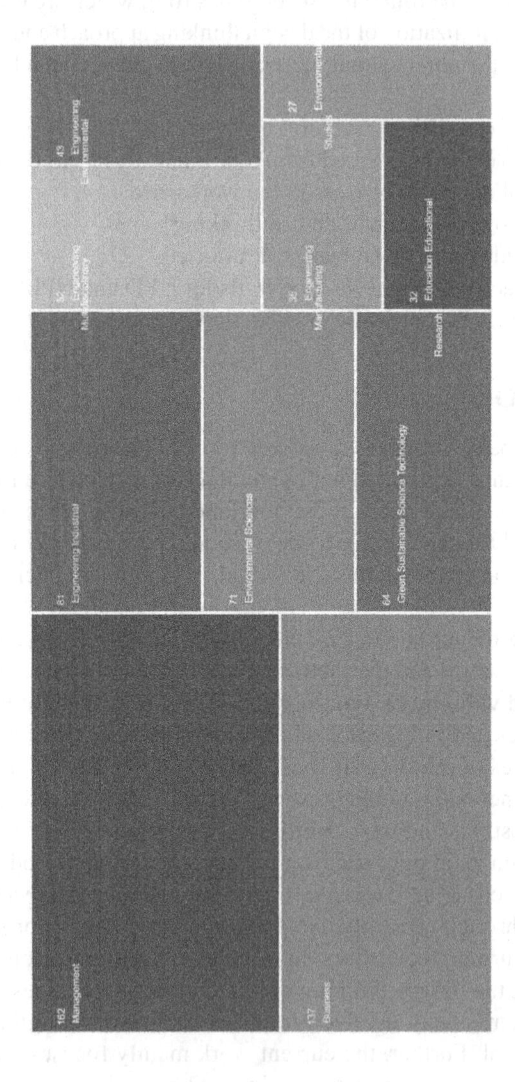

FIGURE 3.8 Design thinking approaches and utilization in various domains [73].

3.7 FUTURE SCOPES

The future scope of design thinking is all about landing ourselves in such a scenario that seems to exist beyond all problems of today. However, the basic design thinking principles, viz. Empathy, Defining, Ideate, and Solutions, remain the same. More study is needed to empirically assess the linkages between design thinking techniques, their practices, brand relevance, and brand equity across time. A design-thinking-based scale system may be incorporated that may give the possible help to clarify the product construction and quantify the relationship of design thinking between product "brand equity and long-term brand competitiveness". Further, many product development processes take the IODA route for developing NPD and NPI. However, in the present scenario, some companies are taking the OIDA route for getting commercial success. This implies that the design approaches are closely related to the types of products intended to be manufactured. One interesting fact is that many of these companies are electronic goods manufacturers, and they possess the immense utilization of state-of-the-art technologies.

Such technology generalization is not limited to the manufacturing of electronics products, but it can also be extended to mechanical products also which is possible only if we advance such technologies to mechanical systems. However, this future of technological shift will bring the revolution in the frequent utilization of design approaches. However, product classification for different product types is the necessity for such technological shifting. More classification and exploration of design factors are necessary for the undergoing trial of design approaches. The traditional ways of thinking of a company's design team or the company's management style are responsible for influencing the design approaches. As a result, they will experience various impacts on the outcome of different design processes. Product types will also define these factors, which will be selectable based on product kinds and product development methods.

Further, using DT approaches, the various prototypes of a product can also be developed, and the DT methodologies will continue to refine the product iteratively based on the user's experience. But this approach of getting feedback from the user and subsequently refining the product is a difficult task, and economic considerations are necessary.

REFERENCES

1. C. Nakata, and J. Hwang, Design thinking for innovation: Composition, consequence, and contingency, *Journal of Business Research*, 118, 2020, pp. 117–128, https://doi.org/10.1016/j.jbusres.2020.06.038
2. M. Gruber, N. de Leon, G. George, and P. Thompson, From the editors: Managing by design, *Academy of Management Journal*, 58, pp. 1–7, https://doi.org/10.5465/amj.2015.4001
3. G. Castellion, and S. Markham, New product failure rates: Influence of argumentum ad populum and self-interest, *Journal of Product Innovation Management*, 30, 2013, pp. 976–979, https://doi.org/10.1111/j.1540-5885.2012.01009.x

4. T. Brown, Design thinking. *Harvard Business Review*, 86, 2008, pp. 85–92, https://hbr.org/2008/06/design-thinking
5. T. Kelly, and D. Kelly, *Creative Confidence: Unleashing the Creative Potential within us All*, New York: Crown Publishing, 2013
6. B. Cousins, Design thinking: Organizational learning in VUCA environments, *Academy of Strategic Management Journal*, 17(2), 2018, pp. 1–18, www.abacademies.org/articles/design-thinking-organizational-learning-in-vuca-environments-7117.html
7. J. Liedtka, Perspective: Linking design thinking with innovation outcomes through cognitive bias reduction, *Journal of Product Innovation Management*, 32, 2015, pp. 925–938, https://doi.org/10.1111/jpim.12163
8. L. Carlgren, M. Elmquist, and I. Rauth, Design thinking: Exploring values and effects from an innovation capability perspective, *Design Journal*, 17, 2014, pp. 403–424, https://doi.org/10.2752/175630614X13982745783000
9. P. Micheli, S. J. S. Wilner, S. H. Bhatti, M. Mura, and M. B. Beverland, Doing design thinking: Conceptual review, synthesis, and research agenda, *Journal of Product Innovation Management*, 36, 2019, pp. 124–148, https://doi.org/10.1111/jpim.12466
10. M. B. Beverland, S. S. Wilner, and P. Micheli, Reconciling the tension between consistency and relevance: Design thinking as a mechanism for brand ambidexterity, *Journal of the Academy of Marketing Science*, 43, 2015, pp. 589–609, https://doi.org/10.1007/s11747-015-0443-8
11. D. Cooperrider, Managing-as-designing in an era of massive innovation, *The Journal of Corporate Citizenship*, 37, 2010, pp. 24–33
12. K. D. Elsbach, and I. Stigliani, Design thinking and organizational culture: A review and framework for future research, *Journal of Management*, 44, 2018, pp. 2274–2306, https://doi.org/10.1177/0149206317744252
13. D. L. Zheng, Design thinking is ambidextrous, *Management Decision*, 56, 2018, pp. 736–756, https://doi.org/10.1108/MD-04-2017-0295
14. R. B. Bouncken, V. Fredrich, P. Ritala, and S. Kraus, Coopetition in new product development alliances: Advantages and tensions for incremental and radical innovation, *British Journal of Management*, 29, 2018, pp. 391–410, https://doi.org/10.1111/1467–8551.12213
15. J. Behrens, and H. Patzelt, Incentives, resources, and combinations of innovation radicalness and innovation speed, *British Journal of Management*, 29, 2018, pp. 691–711, https://doi.org/10.1111/1467–8551.12265
16. T. Morgan, M. Obal, and S. Anokhin, Customer participation and new product performance: Towards the understanding of the mechanisms and key contingencies, *Research Policy*, 47, 2018, pp. 498–510, https://doi.org/10.1016/j.respol.2018.01.005
17. P. Frow, S. Nenonen, A. Payne, and K. Storbacka, Managing co-creation design: A strategic approach to innovation, *British Journal of Management*, 26, 2015, pp. 463–483, https://doi.org/10.1111/1467-8551.12087
18. B. R. Martin, The evolution of science policy and innovation studies, *Research Policy*, 41, 2012, pp. 1219–1239, https://doi.org/10.1016/j.respol.2012.03.012
19. R. B. Bagno, M. S. Salerno, and D. Oliveira da Silva, Models with graphical representation for innovation management: A literature review, *R&D Management*, 47, 2017, pp. 637–653, https://doi.org/10.1111/radm.12254
20. H. Evanschitzky, M. Eisend, R. J. Calantone, and Y. Jiang, Success factors of product innovation: An updated meta-analysis, *Journal of Product Innovation Management*, 29(S1), 2012, pp. 21–37, https://doi.org/10.1111/j.1540-5885.2012.00964.x

21. D. H. Henard, and D. M. Szymanski, Why some new products are more successful than others, *Journal of Marketing Research*, 38, 2001, pp. 362–375, https://doi.org/10.1509/jmkr.38.3.362.18861
22. K. Kim, and K. P. Lee, Two types of design approaches regarding industrial design and engineering design in product design, 11th International Design Conference—DESIGN 2010, Industrial Design, 2010, pp. 1795–1806
23. www.scopus.com/term/analyzer.uri?sid=47a0b894f95a545a344ae980fa01d9ca&origin=resultslist&src=s&s=TITLE-ABS-KEY%28design+AND+thinking+and+product+innovation%29&sort=plf-f&sdt=b&sot=b&sl=57&count=1237&analyzeResults=Analyze+results&txGid=d6bf2d73d7170d5d9f9693b9d6a47013 (Accessed on: 1–24–2022)
24. T. Brown, Design thinking, *Harvard Business Review*, 86, 2008, pp. 84–92, https://hbr.org/2008/06/design-thinking
25. T. Brown, *Change by Design: How Design Thinking Can Transform Organizations and Inspire Innovation*, 1st ed., Harper Collins, New York, 2009
26. M. Geissdoerfer, N. M. P. Bocken, and E. J. Hultink, Design thinking to enhance the sustainable business modeling process, *Journal of Cleaner Production*, 135, 2016, pp. 1218–1232, https://doi.org/10.1016/j.jclepro.2016.07.020
27. H. Plattner, C. Meinel, and U. Weinberg, *Design Thinking: Understand—Improve—Apply*, Springer, Berlin, Heidelberg, 2009, https://doi.org/10.1007/978-3-642-13757-0
28. V. P. Seidel, and S. K. Fixson, Adopting design thinking in novice multidisciplinary teams: The application and limits of design methods and reflexive practices, *Journal of Product Innovation Management*, 30, 2013, pp. 19–33, https://doi.org/10.1111/jpim.12061
29. B. Leavy, Design thinking—a new mental model of value innovation, *Strategy and Leadership*, 38, 2010, pp. 5–14, 10.1108/10878571011042050
30. D. Dunne, and R. Martin, Design thinking and how it will change management education: An interview and discussion, *Academy of Management Learning & Education*, 5(4), 2006, pp. 512–523, https://doi.org/10.5465/AMLE.2006.23473212
31. M. Razavian, A. Tang, R. Capilla, and P. Lago, In two minds: How reflections influence software design thinking, *Journal of Software: Evolution and Process*, 28(6), 2016, pp. 394–426, https://doi.org/10.1002/smr.1776
32. M. Holloway, How tangible is your strategy? How design thinking can turn your strategy into reality, *Journal of Business Strategy*, 30(2/3), 2009, pp. 50–56, https://doi.org/10.1108/02756660910942463
33. L. Carlgren, I. Rauth, and M. Elmquist, Framing design thinking: The concept in idea and enactment, *Creativity and Innovation Management*, 25(1), 2016b, pp. 38–57, https://doi.org/10.1111/caim.12153
34. R. Buchanan, Wicked problems in design thinking, *Design Issues*, 8(2), 1992, pp. 5–21, www.jstor.org/stable/1511637?origin=JSTOR-pdf
35. D. A. Schön, Educating the reflective practitioner, *Jossey-Bass Higher Education Series*, 63(1), 1987, pp. 51–61
36. S. Chen, R. Benedicktus, Y. Kim, and E. Shih, Teaching design thinking in marketing: Linking product design and marketing strategy in a product development class, *Journal of Marketing Education*, 40, 2018, pp. 176–187, https://doi.org/10.1177/0273475317753678
37. U. Johansson-Skoldberg, J. Woodilla, and M. Cetinkaya, Design thinking: Past, present, and possible futures, *Creativity and Innovation Management*, 22, 2013, pp. 121–146, https://doi.org/10.1111/caim.12023

38. W. Brenner, F. Uebernickel, and T. Abrell, Design thinking as mindset, process, and toolbox: Experiences from research and teaching at the University of St. Gallen, *Design Thinking for Innovation*, Switzerland, Springer, 2016, pp. 3–21, https://doi.org/10.1007/978-3-319-26100-3_1
39. M. Crawford, and A. Di Benedetto, New Product Management, 12th ed., McGraw Hill, New York, 2020, www.mheducation.com/highered/category.10383.new-product-management.html?page=1&sortby=relevance&order=desc&bu=he
40. J. Kolko, Abductive thinking and sensemaking: The drivers of design synthesis, *Design Issues*, 26(1), 2010, pp. 15–28, 10.1162/desi.2010.26.1.15
41. K. Michlewski, Uncovering design attitude: Inside the culture of designers, *Organization Studies*, 29(3), 2008, pp. 373–392, https://doi.org/10.1177/0170840607088019
42. R. G. Cooper, The dimensions of industrial new product success and failure, *Journal of Marketing*, 43, 1979, pp. 93–103, https://doi.org/10.2307/1250151
43. S. L Beckman and M. Barry, Innovation as a learning process: Embedding design thinking, *California Management Review*, 50(1), 2007, pp. 25–56, 10.2307/41166415
44. B. Sandberg, and L. Aarikka-Stenroos, What makes it so difficult? A systematic review on barriers to radical innovation, *Industrial Marketing Management*, 43, 2014, pp. 1293–1305, https://doi.org/10.1016/j.indmarman.2014.08.003
45. 5 Stages in the Design Thinking Process, Interaction Design Foundation, 2020 www.interaction-design.org/literature/article/5-stages-in-the-design-thinking-process (Accessed on: 3–27–2020)
46. J. Liedtka, T. Ogilvie, and R. Brozenske, *The Designing for Growth Field Book*, Columbia University Press, New York, 2019
47. H. E. Lin, and E. F. McDonough, Cognitive frames, learning mechanisms, and innovation ambidexterity, *Journal of Product Innovation Management*, 31, 2014, pp. 170–188, https://doi.org/10.1111/jpim.12199
48. C. Wu, S. K. Parker, and J. P. J. de Jong, Need for cognition as an antecedent of individual innovation behavior, *Journal of Management*, 40, 2014, pp. 1511–1534, https://doi.org/10.1177/0149206311429862
49. https://uxplanet.org/principles-of-design-thinking-stages-of-design-thinking-b2cc219063ac (Accessed on: 2–2–2022)
50. www.thinkwithgoogle.com/intl/en-apac/future-of-marketing/creativity/design-thinking-principles/#:~:text=Design%20thinking%20builds%20people's%20capacity,and%20practice%20a%20certain%20mindset.&text=The%20next%20time%20you%20need,%2C%20expansive%20thinking%2C%20and%20experimentation (Accessed on: 2–2–2022)
51. www.interaction-design.org/literature/article/5-stages-in-the-design-thinking-process (Accessed on: 2–2–2022)
52. Y. Haik, *Engineering Design Process*, Thomson Learning, Pacific Grove, CA, 2003
53. D. G. Ullman, *The Mechanical Design Process*, McGraw-Hill, Singapore, 2004
54. N. F. M. Roozenburg, and J. Eekels, *Product Design: Fundamentals and Methods*, John Wiley & Sons Ltd, Chichester, UK, 1995
55. I. Horváth, A treatise on order in engineering design research, *Research in Engineering Design*, 15, 2004, pp 155–181, https://doi.org/10.1007/s00163-004-0052-x
56. K. T. Ulrich, and S. D. Eppinger, *Product Design and Development*, McGraw-Hill, Singapore, 2008
57. G. Pahl, W. Beitz, J. Feldhusen, and K. H. Grote, *Engineering Design: A Systematic Approach*, 3rd ed., Springer-Verlag, London, 2007, ISBN: 978-1-84628-318-5
58. J. Eekels, and N. F. Roozenburg, A methodological comparison of the structures of scientific research and engineering design: Their similarities and differences, *Design Studies*, 12(4), 1991, pp. 197–203, https://doi.org/10.1016/0142-694X(91)90031-Q

59. R. Kumar and K. Kumar, *An Intelligence Assisted Cobots in Smart Manufacturing, Advanced Computational Methods in Mechanical and Materials Engineering*, Chapter 2, 1st ed., CRC Press, Boca Raton, 2021
60. Martin, R. The innovation catalysts, *Harvard Business Review*, 89, 2011, pp. 82–87
61. https://blog.experiencepoint.com/airbus-design-thinking-leads-to-new-solutions (Accessed on: 4–2–2022)
62. www.reutersevents.com/pharma/commercial/design-thinking-human-centered-approach-pharma (Accessed on: 2–2–2022)
63. www.ideo.com/post/design-thinking-for-educators (Accessed on: 2–2–2022)
64. T. Brown, *Change by Design: How Design Thinking Transforms Organizations and Inspires Innovation*, HarperBusiness, New York, 2009
65. T. Brown, and J. Wyatt, design thinking for social innovation, *Stanford Social Innovation Review*, Winter, 2010, pp. 30–35, https://designthinking.ideo.com/resources/design-thinking-for-social-innovation
66. I. Rauth, L. Carlgren, and M. Elmquist, Making it happen: Legitimizing design thinking in large organizations, *Design Management Journal*, 9, 2014, pp. 47–60, https://doi.org/10.1111/dmj.12015
67. I. Rauth, *Understanding Management Ideas: The Development of Interpretability*, PhD thesis, Chalmers University, Gothenburg, 2016
68. L. Carlgren, M. Elmquist, and I. Rauth, The challenges of using design thinking in industry—Experiences from five large firms: The challenges of using DT in industry. *Creativity and Innovation Management*, 25(3), 2016, pp. 344–362, https://doi.org/10.1111/caim.12176
69. M. Assink, Inhibitors of disruptive innovation capability: A conceptual model, *European Journal of Innovation Management*, 9(2), 2006, pp. 215–233, https://doi.org/10.1108/14601060610663587
70. J. Bessant, C. Öberg, and A. Trifilova, Framing problems in radical innovation, *Industrial Marketing Management*, 43(8), 2014, pp. 1284–1292, https://doi.org/10.1016/j.indmarman.2014.09.003
71. V. M. Story, K. Daniels, J. Zolkiewski, and A. R. Dainty, The barriers and consequences of radical innovations: Introduction to the issue, *Industrial Marketing Management*, 43, 2014, pp. 1271–1277, https://doi.org/10.1016/j.indmarman.2014.09.001
72. H. Luotola, M. Hellstrom, M. Gustafsson, and O. Perminova-Harikoski, Embracing uncertainty in value-based selling by means of design thinking, *Industrial Marketing Management*, 65, 2017, pp. 59–75
73. www.webofscience.com/wos/woscc/analyze-results/57b669b1-ab25-4954-805f-ad1cb2cbcff1-21cc5350 (Accessed on: 1–24–2022)

4 Design Thinking—The Modern Approach in AIML

Mridula G. Narang and Supriya M. S.
M. S. Ramaiah University of Applied Sciences, Bangalore, India

CONTENTS

4.1 INTRODUCTION

Design process always involves design thinking and its support tools to idealize and express design preparations. To obtain appropriate solutions using these design innovative methods, further elaboration of design thinking methods or ideas are required. These innovative ways also need to be powerful enough so that they can balance between conceptual model's expressive actual innovation

DOI: 10.1201/9781003189923-6

and operational framework (Walch and Karagiannis, 2020). It is important to achieve these goals with the help of techniques like Artificial Intelligence (AI) and Machine Learning. While using these techniques appropriately, it is necessary to assess parameters like design methods, improvements in application, and information and data required so that they form a proper integrated approach. The main objective is to study and understand conversion in innovative practices. To achieve these types of goals, AIML techniques needs to create particular application understanding in a required environment with technological knowledge from engineering in an operational environment (Akerkar, 2019).

AI requires structured and unstructured data. This data is then processed, and, with the help of main AI computation tools to create environment for problem solving or reasoning, Machine Learning knowledge base is required for value-added information. Broadly, AI can be defined as the ability of a machine to execute required tasks related to the human way of thinking, such as identifying, thinking, acquiring knowledge, acting together with the environment, understanding, and resolving problems with creativity (Kietzmann and Pitt, 2020)

AI can be used more effectively to achieve aforementioned goals in the fields like robotics where it can be used to take a decision on the basis of a particular environment, to idealize an environment for a machine to have a respective vision property, to process language so that interaction becomes easy; and machine should be able to learn or update the process by learning or updating rules required to process (Guzman and Lewis, 2020)

AI is giving a new platform by finding suitable solutions with the help of current advance technology, and it is widely used in many subdivisions across different industries. Effectively, AI can be used to carry out/track many secure transactions or facilitate in difficult/complicated decisions where it can affect financially in case of asset allocations and wealth management. AI helps to increase efficiency and time management, assess market situations, and find new openings for industries like customer relations administrations and sales administration. To support AI, the Machine Learning tool can be used for its applications, for use in algorithms, and for meeting its other challenges. In order to make Machine Learning a proper development tool in any required service needs required accuracy and interpretability so that the managers benefit from all these techniques. This approach can be an important part in the innovation process. AI plays an important part in the specification of critical goals, data gathering, and formulation with modelling as well as profit/benefit addition (Akerkar, 2019).

Machine Learning (ML) is a powerful tool which helps us to address problems and questions that were not possible to face with traditional programming. Wide use of ML system and its similar features creates additional benefits for design thinking and human-centered activities (Rizzo et al.).

4.2 ARTIFICIAL INTELLIGENCE

Artificial Intelligence is nothing but using science and engineering in making a machine do simple tasks or achieve goals as we humans do. Humans use their intelligence to do these tasks, whereas these machines use computer programs to complete these tasks. Intelligence is the computational portion of the capacity to accomplish goals. Part of our intelligence involves mechanisms hence computer programs or AI can give us very beneficial results by doing these tasks effectively. AI entails studying the problem and identifying solutions that are not obvious while searching for a solution or necessitate a significant amount of computation effort. Few branches of AI are Machine Learning, Deep Learning, Natural Language Processing, Fuzzy Logic, Expert Systems, Genetic Programming, and so on. Widely used few applications of AI are Online Shopping, Marketing, Social Media, Banks, Healthcare, Enhanced Image, Smart Cars, Surveillance, Gaming, and so on. Machine Learning is an approach to design intelligent systems, which adopt its behavior based on data. Growth in the commercial use of AI is due to success of Machine Learning algorithms.

Regarding AI research, a promising decade was 1960s when Simon and Newell's work in specifying an algorithm for problem solving stood out. They used both computer and human techniques to solve problems. Because computers were still in their early stages at the time, resources such as speed and memory to support the requirements of these developed algorithms were not available. This algorithm was correct theoretically, but its supporting hardware were not available. Zadeh's introduction of fuzzy logic and sets, which later became the AI branch of fuzzy logic, was another notable development in AI. Important basis of his work was that even computers can effectively work with fuzzy logic which was then a precise and discrete pattern. In general, human brain working functions were not available at that time as human brains could not be represented in an effective way by computers. People have gradually come to believe that computers that simulate the human brain are capable of making decisions on their own. In the same period, important work was carried out in the form of communications by replacing computer coding by the usage of natural languages. Program developed by Joseph Weisenbaum was named as ELIZA, which made some users feel virtually that they were speaking with a human rather than a machine (Norris, 2017).

In the 1970s, AI was slow in development as the progress in computing technology was growing slowly. In this period, computers available to researchers were limited in number, which were able to perform these kinds of tasks. But a substantial development happened in the fields of natural language processing, image analysis, and recognition.

In 1980s, significant development happened in AI due to the inception of PC. The development of expert systems started, and it started to lead the way to applications of AI in different sectors like business and industrial/manufacturing. At the same time, similar to AI, many researches were carried out in fields such as robotics and expansion of real robots. Also, modern computing in AI started to develop with the help of Moore's law, which was an unbelievable improvement in computer performance which was most required for AI growth and power. The invention of more transistors in integrated circuits resulted in an increase in the number of computers produced per year. Many remarkable developments were achieved in 1990s. In the year 1997, a remarkable achievement was a win over chess expert Garry Kasparov by computer system of IBM deep blue.

After these innovations, the computer machines started to perform "intelligently", but a common definition of intelligence is complicated. In many dictionaries, it is defined to be the learning aptitude and value or to overcome new or trying conditions, reasons or the expert ways of reasoning or the capacity to utilize wisdom to operate one's nature or to think conceptually as measured by neutral conditions or mental sharpness (Norris, 2017). Human intelligence is located in our brains, but it is controlled by our bodies through sensory systems such as taste, touch, vision, smell, and hearing, all of which contributes to our intelligence levels. The human body possesses remarkable talent to compensate when one or more of the sensory systems become particularly injured or destroyed. Humans have abilities that allow them to understand and discover their true nature. This, in turn, leads to the development of experience as well as knowledge. These machines, however intelligent, cannot be compared to replicate human intelligence (Akerkar, 2019).

Day by day, AI is increasingly becoming a part of our everyday lives due to its recent advance in deep learning applications from unstructured and unlabeled data. The meaningful learning aptitudes are further accurate as well as efficient because of extraordinary approach to information and improved computing control. AI promises to meaningfully transform the existing business models and at the same time create new ones. Scalability, longevity, continuous improvement capabilities, increase in productivity, lower costs, and reduced human errors have made AI appear more advantageous than human intelligence.

4.2.1 Benefits of AI (Akerkar, 2019)

AI has benefits in most of the sectors, this cutting-edge technology enables machines to act autonomously, resulting in the efficient execution of iterative tasks. Some of the benefits of using AI are listed below:

(a) Finding out required meaningful and/or useful patterns in large volumes of data of any type, which may be text, video, images, or other unstructured data.
(b) Building improved decisions more rapidly by maximizing the value of all data, moving from analytical questioning to rigid questioning.
(c) Self-learning models allow you to adjust quickly to fluctuations in the patterns of data and underlying business situations.
(d) Makes it easier for business users and data scientists to understand, respond to, and use data through better visualization and transparency.
(e) Recognizes unique understanding about models in the data, which allow optimal customer separation and behaviors.
(f) Offers new business prototypes and value creation by fast-tracking innovation through finding new forms in data and fully applying knowledge assets.

4.2.2 Advantages of AI in Various Sectors

Customer relationship management (CRM)— the relationship with the customer makes marketing efforts real and operative. Hence, an importance is always given to improve relationship with customer. This also can be achieved with the help of AI. Regression analysis and clustering techniques from AI can be used in such a way that a customer of a specific company or a service used by different people can become associates by studying their statistical records and changes in lifecycle.

Generally projecting analytics is used if there is a need to locate outliers, fraud, or scams. In this, emphasis is given on to find essential designs, differences revealing things that are different. In the financial sector, it is widely used to detect scam. Pharmaceutical, medical, and many other sectors also use same techniques.

Forecasting demand—For all businesses, it becomes a vital and inspiring assignment to project a demand of new products, change design in a product, and adding new services. With the time-series data, a wide-ranging estimate was made, but now it is possible to predict sales of a product at the lowest level by preserving all of a client's information.

Refining processes for different sectors—In many sectors, the use of a multipart and sensitive machinery has become common. Stoppage of these machinery leads to idle time, or complete shutdown can happen which will lead to a disturbance or affect production time. Hence, it is necessary to keep a track of machine part replacements or maintenance shutdown periods by its old records and performance data, comparing old data and real-time data to improve the functioning and ideal time of machine. Also, it can be used to alert before any unexpected machine breakdown happens. Development in the reference engines—Many services are available online today, and in order to use these services, the available service must be dependable. This can be achieved by using different AI techniques depending upon requirement (Akerkar, 2019).

Any business's human resource department can use historical data to improve the hiring process, find a suitable candidate, or retain existing candidates based on their previous performance and general behavior, predict dissatisfaction, and need to lead a discussion to retain a leaving candidate using AI.

4.3 MACHINE LEARNING

In 1950s, a concept of ML started to develop. To develop ML was to bring advancements in prediction accuracy through advancements in existing methods and using different approaches that would support more advancement in performance and work with its tools to provide a better visualization by keeping different approaches on available data or most useful data. An important feature in ML algorithm is to improve performance by learning and using effectively acquired knowledge than coding more stringent rules. Generally, we can define ML problems as being controlled, uncontrolled, and reinforced. When we use ML to forecast the problems on the set patterns, they can be considered to be controlled. Similarly, when we use an associate data, for example, we use a particular data to consolidate and understand effectively. The difference between the two is that the latter uses old data to analyze or predict the future, whereas the need is to make the machine to use its own intelligence to make predictions in this case. In the last set of problems, our objective is to learn a set of activities to achieve a specific task or desired outcomes with the help of ML to find a most effective way to use reinforced hints (Norris, 2017).

Any work related to data and knowledge such as the removal or insertion in existing data forms a subject related to data science, but using ML effectively processes data which will lead to efficiency in business. ML techniques can be used in four ranges that are grouping or sorting, collecting, involvement learning, and numeric forecasting. Grouping and sorting out data are related to sorting of text from provided documents or to sort out from well-defined sets and by separating text or information from other sources. As per the requirement, it can be used in different document text/paragraphs/sentences or words, etc. In collecting techniques generally, analysis is done on the available data to find groups of information which is similar with the help of existing grouping procedures on data sets. It becomes more difficult in the case of identification of text or sentence or a word for grouping by applying a particular procedure or method. To achieve the goal of identification and sorting of data, there is a need to develop an interaction with subjective data or other features will lead to many new insights which can be used effectively to achieve a goal (Rajendra Akerkar, 2019).

In numeric forecasting, a soaring or grouping of a continuous data needs a totally different approach to sort, group, and to repeat the procedures. In ML, the main focus is on to find out a set of important information with a particular procedure. This is accomplished through the continuous use of modifying data, revising data, and updating data with changing environments to provide information that is more valuable in achieving a goal with greater accuracy. This is the most important achievement by developing a set of data that can give us new directions or new patterns, even if it is applied in the case of a changing environment or when there are changes in information on its own, simply by collecting very few additional data rather than going for changes in programmes or procedures. Because of these features and compatibility of ML, it has become a prominent tool in AI and is widely used in many processes and forms a part of machine understanding processes such as language processing, image identification or image generation, and identification of music and texts and updates the knowledge so that it can predict a changing environment on its own (Rajendra Akerkar, 2019).

In the ML process, we must find an approach that allows machines to learn and apply a set of data on their own to generate a specific functional requirement and to use a revised procedure appropriate to the current environment, or we must use both approaches to maximize and enhance the approach as per specific product or assignment. For example, in the case of online shopping, generally it will help to predict future requirements as per the list of old purchased products or the usage of service by a customer or to suggest a new product which has modifications related to old purchases or services. Similarly, in transport business, a major help can be achieved in case of cost or benefits by selecting an optimized route or to suggest the most beneficial route.

4.3.1 List of Activities Required for Machine Learning to Meet Required Objectives

The most important feature of ML is that there is no involvement of any personal interaction or change in approach to a data record with respect to the required application itself by adopting processing skills of data which makes these machines capable enough to decide or adopt a certain procedure to take a decision rather than wait for new data inputs and then taking a decision based on related data provided. Also, a modification or iteration on data comes in to effect as soon as it becomes old or is not much related to new environment or is in the need of a higher accurate prediction.

The list of activities are used as follows (Akerkar, 2019):

(a) Data Collection: Collection of maximum suitable records and merging them into a dataset
(b) Data Preparation: The main aim is to process data in a practical way by filling the gaps or in the case of unavailability of a particular set of data. Procedures for data preprocessing and cleaning should be advanced.
(c) Data Divisions: Breaking a specific set of detached data for a procedure to gather new information in the existing data so that a proper goal can be easily achieved; additionally, a new goal can be fulfilled through iteration.
(d) Guide or Development of a Suitable Process: Application of a particular portion of data to evaluate and help to establish a procedure that gives various information to predict the best outcome or required optimum solution.
(e) Test and Confirm a Developed Process: The objective of a developing process is that it should forecast accurately with the help of available data. By checking this on different situations of available data, we will come to know the accuracy of the developed process. Hence, before applying to any actual data we will come to know whether it requires any modification or not.
(f) Operation on developed process: The aforementioned checked developed process can be used as the main process involved in order to carry out an optimized solution in a given set of data.
(g) Repeat: Repeat the aforementioned developed process again on the data (either newly formed or data which is revised) as per requirement or situation to increase a percentage of accuracy.
(h) ML Productivity: To make ML more productive, it is necessary to use a particular set of processes and use a refined old data effectively.

4.3.2 Benefits of ML in Different Sectors

The most common functional sets are as follows (Dam and Siang, 2018):

(a) Customer Investigation: This category is mostly used by a marketing segment, for example, to find out a group of people using similar search or having similar social contacts or as per particular brand or as per cost or their old experience, etc.
(b) Supply Chain Investigation: In this module, it includes more usage related to supply material chains in terms of predicting the need of a product, its availability, its price-fluctuation-related issues, storage issues, cost effectiveness due to storage, etc. Similarly this segment involves more important resources like humans as well as other reserves in production.
(c) Investigative Fraud or Analysis Based on Risk Involved: Generally this is used in insurance or financial agencies for example credit cards limits for particular set of customers, approving claims in insurance companies, change of requirement in the market or operating in different financial risks.
(d) Logical Usage in Government for General Public: This method can be used by the government in many critical situations—for example, when there is a need to fix a route of new pipeline, to detect and find possible ways to repair and refunctioning of existing distribution system, to divert additional resources to less available resource areas, to transportation systems, and many others.

4.3.3 Critical Comparison of ML

Use of ML can be broadly classified in three categories. The first category is informative investigative, and this can used for collecting data in the form of strong points of a particular sector and compiled predefined set of information where ML needs to apply. Second category that is the application or usage of ML is used to predict future and is known as projecting investigative whereas the third one is used to get the best outcome after the application of ML tools. For example, if ML is used for business sector, then the following points need to be considered (Akerkar, 2019).

(a) Gathering Business Information of All Areas: In this, the main objective is to collect or do development in a process where a knowledgeable decision can be taken with collective inputs or information from all areas related to its performance continuously during the course of the business
(b) Understanding: Acquiring complete perspective of customers with respect to business deal history, dealing in divisions, responses, beliefs, etc., to take effective decisions and assist in development with respect to profits in a particular business.

(c) Foreseeing: Forecasting to calculate prospective results by considering old data or old actions that are taken to increase profitability or to achieve sale targets or improved design of a product, etc., which will lead to generating more business.
(d) Business Responsiveness: Using a powerful optimized technique in both people-oriented methods and program-oriented methods.
(e) Planned Position: Align different activities in a business right from planning stage to that of implementation in such a way that it may help in keeping the originality and a wider range of operations. Decision-making in a particular business is related to its favorites, urgencies, goals, necessities, and so on, which help in the decision-making process. Data and information can be used to calculate strength. Before launching the application, it is necessary to investigate each and every set of data pertaining to available systems, asset scheduling, making a customer an associate, time and availability of required resources in the activity, business with the aid of machines, bonus, or any benefits transferred to customers. Once this data is properly identified, it can be effectively introduced in a process. This data can be structured and unstructured in nature. By processing this data, complete knowledge can be gained for taking a required optimum decision. Suitable data for decision-making with similar agreements, satisfaction reports about product or a service, change in approach as per the requirement of customers, collecting data from other resources as per their need such as the availability of vehicles for transport or availability of human personal, segregation of data collected from different social networks, and others (Akerkar, 2019).

4.4 INFORMATIVE INVESTIGATIVE

Knowing how a business is run by using past records and information gathered from old events for data provides us with the necessary foundation to assess the situation and determine the need for an appropriate procedure adoption link. In this investigation, we will be able to get well-defined past data like old sales records, profits earned in past years, costing required to generate the required profits, list of resources required to complete the tasks, and critical resources and also we can get similar data for the actual period (Akerkar, 2019).

4.4.1 Projecting Investigative

In this investigation model, the main focus is given to predicting future events with the help of refined search engines and with the help of higher-level investigation on old or new structured data and mathematical algorithms. To project an event, there is a need to develop a proper link of inputs provided or revised in a set of data and the generation of output. After this process, a repetition of the same is carried out to achieve a more accurate output. A process of predicting events will start with the investigation of old data and get updated to a new set of information by assessing, sorting, and grouping, etc., repeatedly with fast iteration procedures. Then, a required procedure is performed to give the required flexibility to get a desired result. When this informative investigation completes the aforementioned step, it passes through a procedure which will detect other changing patterns or evaluate the new information for the modified data to predict future which will depend on change in behavior and latest trends etc. as these things affect the accuracy of prediction. The projecting investigative model is used on past data as well as on generated new data with respect to time and developing subsequent relationships with time and modified data, as a result of which an understanding of repetitive sets of data occurrence and describing the changeability of new data is gained. For example, this is effectively used in the case of replacement of parts of machine in manufacturing units considering usage rate and known, predicted changes in parts to fulfill manufacturing needs (Akerkar, 2019).

To predict the future, it needs a data of normal behavior, but if these data are not exhibiting normal behavior or are fluctuating then there is a need to identify this kind of data and to apply required procedures to predict the future with systematic action plans. This kind of projecting investigation can be applied to many sectors so that their goals can be easily and effectively achieved. For example, in the insurance sector, this method can be effectively used to detect or investigate new claims with the help of old records where there is more probability of providing untrue or false information. Similarly, it can be applied to get insights on most faithful policy holders, to focus on a specific list of policy holders so that more business can be carried out, to find out how people will react with particular treatment and cost related to such treatments, helpful in comparing and approving or declining claims etc. Similarly, it can be used in medical sector to identify people who are vulnerable to different diseases and for other reasons. In the case of health-related sectors, it is possible to track a condition of people having a specific disease, which may be a trigger to hospital regarding the need of a particular service well in advance. In the financial segment, it is much more useful to detect and track different cases of financial transactions or statements or activities or predicting by spending relatively very less cost. It can also prove to be useful in the case of sectors involved in services providing business, targeting particular segment for sales or predicting customers for particular service and help the marketing division to achieve targets and to announce different schemes to keep customer base wide and intact.

Projecting investigative method is applied so that it becomes more feasible to achieve required goals. In the case of financial sectors, a lot of issues can be resolved with the help of developed different networks to predict financial transactions, relating the past records, locating new opportunities or scope for new customers and to understand the capacity of a customer to repay loans that will help to take a decision regarding fresh loan approvals. Prediction can be done for a set of customers referring to old data to find out a valid set of customers. For predicting the effectiveness of new marketing strategies based on past records of success, there is a need of altering or revising these strategies to give the maximum benefit. The advantage of projecting investigative model is that it works repeatedly on old data and regenerates data so that more accurate or newly required decision-making processes and new insights can be formed, which are necessary to achieve desired goals (Akerkar, 2019).

4.4.2 Regulatory Logical Processes

In this advanced module, different techniques are used to find an optimized solution or replicated solution by controlling different resources to achieve the desired goals or to establish new strategies to avoid competition and adopt different ways to have the best and optimized solution by dealing with the uncertainty of data (Akerkar, 2019).

4.4.3 AI Being Implanted into Business Models

To get the best results by incorporating AI into respective models and achieve maximum benefits every time, you'll need the right skills, such as proper definitions, AI field experts, established procedures to address difficult issues, improvements in critical issues, and a strategy to meet your objectives. By a proper implementation of AI in a business, it is possible to collect and gather required information, process this enormous amount of data to reach different insights, and develop a programmed solution for decision-making or suggest a new approach and a way to get effective optimum solution (Akerkar, 2019).

4.5 DESIGN THINKING

Design thinking is a repetitive procedure. This procedure works toward keeping information regarding the usage done by customer, experiment theories, and redefinition of difficult situations intact to detect the substitute methods of strategies and solutions which cannot be noticeable at the time of initial level of understanding and also it provides solution-based approach to the problems (Dam and Siang, 2018).

4.5.1 The Need to Know Different Ways of Thinking

Generally, as humans we can gain a new or updated set of knowledge by applying our old gained knowledge or experience on the different thinking processes. We are capable of understanding a particular process more effectively by gaining this additional knowledge and then finding different ways of solutions to a given problem. In this process, we can think on the basis of old and new knowledge to find an optimum solution by considering all required aspects. If a problem is similar to an earlier one, then it is possible to predict the solution by the comparative analysis of both the problems. Advantage of this is that it is very fast, and we know what type of results we will get after applying a known solution to a problem. This saves our lot of efforts to predict accuracy, time, and other resources required to find different solutions, to find a new approach, need to understand a problem, and so on. Human mind always compares earlier events with new events and decides on similarities of both the events by identifying them with an earlier gained set of knowledge.

We humans remember a particular object by its noticeable characteristics like shape, color, weight, size, or any other special features. Our mind compares these things with a new object as soon as it comes in the front. Our mind first compares and takes a particular decision about an object. As soon as our mind sees a new object, it starts remembering the old object even though only few things matches with the new object. We block our thinking by replacing thoughts and observations of the new object, which is actually not correct or may lead us to misconceptions regarding the new object and coming to a set of wrong decisions even when all features do not match. Hence, it is very important to check the object individually rather than just comparing only similar or matching characteristics. It needs different approach of thinking to find solutions of a particular problem (Dam and Siang, 2018).

Many incidents occur in our daily lives that may be related to this type of case where our mind compares new problems to old problems in order to find a required solution and fails to recognize that finding an optimum solution by comparing with old experiences is far easier than finding a new solution. Also, we need to understand that our thinking is based on our past experiences rather than finding other ways to find an optimum solution for the same problem.

Once a patient was suffering from ascites, a condition in which fluid is accumulated in the abdomen. As the accumulation of the fluid in the abdomen affects the lungs, kidneys, and other organs of the body, the fluid needs to be removed from the body at the earliest. Thus, the doctors began extracting the fluid from the patient's body. During this procedure, a small puncture wound occurred, and due to this a considerable amount of body fluid was lost. This worried the doctors as excessive loss of body fluid might result in various complications. Various MBBS and MD doctors gathered to analyze the problem. One of them suggested a surgery; another one suggested stitching up the wound; and so on. Meanwhile, a nurse who was observing all these made a suggestion. She suggested the doctors to make the patient lie on the other side, as this would result in the collection of water on the other side and would not leak from the puncture wound. Hearing this, the doctors were surprised and exclaimed as to how they had not thought of this solution before. Hence, we can observe that most of the times the most optimized solution to our problem is right under our nose, but our mind doesn't realize this.

From this aforementioned story, we can understand our general way of thinking process by relating things with old knowledge and experience to arrive at a kind of solution to a problem. This solution may or may not be optimum as well, because it may become harder for us to evaluate other solutions as our thinking process stops as soon as we find a solution of problem. Similarly, it happens for the people who are experts in a particular profession where they apply knowledge and gain experiences for a particular set of problems in the specific field only.

Hence, it becomes very difficult for these people to change their thinking pattern suddenly and find another solution for the same problem. This means that after a particular stage our mind stops the thinking process and follows a monotonous path directly rather than finding a solution to a problem by thinking.

Design thinking is a process in which a program is developed to process and update a massive amount of data every time before suggesting a solution to any given problem. Hence, the most important thing in this is establishing a set of data parameters which we generally use while making a decision to meet all the requirements or expectations necessary for getting solution to a particular problem. As instance, to develop a new product or service provider platform it's possible to make alterations and improvements in the old existing product itself. For instance, mobiles are used in today's life for communication. Let's go back to a period of 5 years. At that time, the main focus was given on communication only. Hence, people started using mobiles more frequently to get the maximum benefit as it was becoming very easy to communicate with anyone from anyplace. People began to demand that these mobile phones be used for more than just communication as time went on. An improvement of technology aided this requirement, and thus now we use mobiles not only for communication but also for several purposes such as playing games; attending virtual classes; to watch movies; to play various genres of music; for our day-to-day activities such as alarms, calculators, and many others; to access various social media platforms such as YouTube, Facebook, Instagram, Snapchat, and many others; and to carry our office related works in free time. In a similar way, we come to a solution to many problems by just keeping a machine to do multiple tasks. Here, a product's demand areas changed from communication to other tasks. Hence, before developing any product, we have to understand the purpose for which it is designed so that the same product can be rendered useful by slight modification for other purposes as well. This can be achieved by interacting more with the product users.

Design thinking becomes more effective by collecting this type of information or inputs from users by processing this information in different ways so that it can be useful to solve other problems also. As design thinking gives us a platform to work on many platforms by changing or by updating data, this can be achieved by outlining different problems and by defining a thinking pattern. After this step, we can check whether we have got an optimized solution or not by the application of this process of design thinking. If we have not achieved an optimized solution then we have to update data once again and work with other old or new possible solutions to the problem so that we get an optimized solution.

What is of essence here is to analyze that the earlier programming approach was to code the information in the computer machine to create applications which would then divide the constraints and develop a relation with the problem so that we get a solution. That is nothing but giving certain rules while processing the data to get the required answers. Now, the same concept is revised, and the application is done in such a way that rather than applying certain set of rules, we are applying different answers while processing the data so that they themselves form as rules while suggesting us different solutions which will be once again processed so that we will get an optimum solution. Generally, we use different programs by applying certain rules to process a set of earlier gathered data from various resources to get answers to the problems. With the help of ML principals, this arrangement is revised as to provide answers with the collected data instead of rules to get rules to apply which will give a solution, and again these solutions are improved by repeating the application process to get a more accurate result (Rizzo et al.). In this approach, we can establish a proper synchronization of data and answers which will deliver optimized solutions to problems. Advantage to develop these kinds of procedure is that a required thinking is based on design thinking and amalgamation with data or knowledge which get updated automatically on its own. Hence, there is no need of any interference of human for inputs, and these programs run on their own learning procedure.

In the development of such kind of programs, different teams of peoples who can logically prepare data also can guide us as experts and a machine to configure in such a way that it will process large quantity of data in short period. The flexibility of design thinking expands our vision, reveals new path, and discovers new approaches. Recently, we designed interactive systems that supported Machine Learning solutions where the design processes were conducted under the umbrella of the Human-Centered Design methods and therefore the mindsets of design thinking. However, methods and mindsets were elaborated and partially transformed to face the new challenges posed by the utilization of the Machine Learning tools. In one case, associated with a system for an early detection of attacks to bank ATMs (Rizzo et al., 2019), the key factor for producing an efficient solution was the identification of a special kind of data source (and the related feature extraction process). The developed methodology allowed us to receive the authorization by the Italian Data Protection Authority to run field experiments with real transactions on ATMs located in city centers. In another case, the winning factor was the live construction of the dataset with the identical operators that use the system for the detection of excessive metal material (burrs) within the forms produced by the Iron and Steel Industry (Rizzo et al., 2019). During this case, the development of the dataset by the collaborative efforts of all the personnel of the corporate led to the chance to find not only the burrs but also problems in the management of pouring. These two case studies presented not only similarities moreover peculiarities as well regarding the methodologies employed, but both needed a revision of the way to approach the design challenges. Resting both on our experience and on well-established Machine Learning solutions led to theories and design methodologies required to be used for creating AI systems that best empower people.

The whole concept started to take shape by starting the development of AI in the year 1969. In this, a five-phase process was suggested by Hasso Plattner Institute of Design at Stanford, i.e., D.school. The principles of design thinking were based on those laid by the Nobel Prize laureate Herbert Simon. D.school is one of pioneer institutes for the development of design thinking process. The five phases of design thinking are as follows:

(a) Understanding the requirement of users.
(b) Defining the problem with respect to users' requirements, problems faced or the need to find solutions to faced problems, and other insights of the required application field.
(c) Updating data and finding different solutions and check.
(d) Trying to start and apply different solutions and redefining the problem.
(e) Testing solutions for optimized solutions.

One should not consider that design thinking can achieve the required result by following mentioned steps only. It can also achieve required results by changing a sequence of occurrences of aforementioned steps. We can consider this as an outline for a process. Design thinking process can help us leading smooth lives smoothly, proving to be very beneficial in many areas of our daily life. We can follow the footsteps of nature and generate a machine which can be used in maximum cases to get an optimized solution as soon as we apply to any given situation or problem.

4.5.2 Role of Design Thinking in Problem Solving Process

The role of design thinking is to develop a method for a product or service which will give us an optimum solution by collecting a set of data regarding customers' requirements, ways, and conditions in which they operate and maximum related data from all the resources, which can be rebuild in such a way that it will be effectively used every time. Consider all the approaches or solutions (they may be newly developed or old existing), reapply them, and check their effect relating to optimizing the solution of a problem. Once we re-create the problem with this kind of approach, we can understand the problem from a different perspective. One more important aspect of this approach is that it will provide us with a complete process to find solutions with required constraints and other features of the problem (Elsbach and Stigliani, 2018)

Don Norman explains that design thinking and its special features are the need of time. If we closely observe, we can find that many people in business or services or in different professions work on problems to find solutions without applying a complete required thought process. The first step is to go for a complete understanding of the problem. Rather than going directly to solve a problem first, we should check for other possible solutions. Once we get all the solutions, one must approach the solution that likely to be an accurate one. Then, we should once again reiterate the process and come to a solution which is optimal. This process is widely used in design thinking (Badke-Schaub et al., 2010).

4.5.3 The Potential of Design Thinking Over Other Approaches

Design thinking processes can keep a project on course and organizing core issues which involves many groups of individuals in many departments; hence developing, reasoning, categorizing, organizing ideas and problem solutions are often difficult. Among the scientific activities, there will be processes analyzing how users interact with products and investigating the conditions under which they operate: researching user needs, pooling experience from previous projects, considering current and future conditions specific to the product, testing the parameters of the problem, and testing the practical application of other problem solutions. Unlike a strictly scientific approach, in which the majority of known qualities, characteristics, and so on of the problem are tested in order to arrive at a problem solution, design thinking investigations include ambiguous elements of the problem in order to reveal previously unknown parameters and uncover alternative strategies. Designers are urged to examine and falsify these problem solutions in order to get at the best available choice for every problem or impediment encountered during the design process. With this in mind, it may be more accurate to say that design thinking is about thinking on the edge of the box rather than outside of it (Dam and Siang, 2018).

4.5.4 Generating Creative Ideas and Solutions by Holistically Understanding Humans

Design thinking aims for a holistic and empathic knowledge of the problems that people experience, with a firm foundation in science and rationality. Design thinking attempts to empathize with humans. This includes ambiguous or intrinsically subjective categories like emotions, needs, motivations, and behavioral drivers. Because of the way design thinking generates ideas and solutions, this method is usually more attentive and is based on the context in which users operate, as well as the challenges and barriers they may experience when interacting with a product. The creative component of design thinking can be found in methodologies that seek to develop problem solutions and insights into the practices, actions, and thoughts of real people.

Dunne and Martin (2006) expound on the design thinking method, which has recently gained traction in organizational settings. This is usually due to the fact that the design of products and services is a significant component of industry competitiveness, to the point that many well-known organizations have committed to becoming design leaders. And, while design thinking has become a vital component of the design and engineering areas, as well as being used in industries, it can also have a good impact on twenty-first-century education across disciplines because it incorporates power in problem solving. As Rotherham and Willingham (2009) point out, pupils in academic settings are expected to read critically, think and reason logically, and solve complicated problems. Thus, in order to help students succeed in this interconnected, digitalized world, educators should assist students in developing and honing twenty-first-century skills (e.g., design thinking, systems thinking, and teamwork skills) that can improve their problem-solving abilities and prepare them for faculty and careers, as Razzouk and Shute (2012) suggests. Innovative design thinking is the need of the hour in business settings, particularly in sectors such as engineering and architecture, because it can revolutionize how people learn and solve problems (Dym et al., 2005). If graduates lack employability skills such as job hunting skills, interview skills, and presentation skills, which demand a certain level of English fluency, success will be a foreign dream. Rifkin (2014) advocates for adopting new approaches to communication as a result of technological improvements in order to achieve holistic growth and development. Furthermore, the impact of Information and Communication Technology (ICT) has advanced within the new design thinking process, as well as the tactics that should be pursued in light of industrial requirements. A good starting point is to employ the topic of graduate preparation in its entirety, which focuses on the needs of current generations without sacrificing the flexibility of future generations. According to Simon, (1996), design thinking is the central or differentiating activity of engineering. It has also been suggested that engineering programs should encourage the development of graduate engineers capable of designing effective solutions to meet social concerns (Evans, 1990). Being successful in today's highly technical and globally competitive environment necessitates the development and use of four skills. Design thinking, like issue solving, may be a natural and pervasive human act. According to Braha and Maimon (1997), a design process begins with being discontent with the current situation and the determination that some action must be made to fix the problem. Throughout their careers, many scientists can design and act as designers, frequently without realizing or being aware that they are functioning during a design process.

In step with McGrew (1970), "The illiterate of the twenty-first century won't be who cannot write and read, but those who cannot learn, unlearn, and relearn". In today's fast changing industrial world, new graduates need to adapt to current situation. As we are exposed to very fast changing environment, it becomes very difficult to find skilled people who can effectively handle these changing environments. There is still a big gap in the distribution of education and skills that are needed for a particular employment. It's a big challenge to address this issue due to which many fields have the shortage of skilled people.

Doepker (2019) predicts "The Future of Jobs" within the World Economic Forum that due to automation many people will lose opportunities as they lack required skills or suitable qualification. There is a requirement that our educational system should work aligning with the needed development in skills and qualifications of students in such a way that they will be able to fill the new vacant positions in this fast-changing environment. Still, there is wide gap in particular skills and qualification in graduates of a particular field. We should be able to address this issue so that graduates can gain maximum benefit out of it to get required employment. To close this gap, experts or related surveys indicate that graduates from university should require training aligning with industry requirement. Graduates required for corporate houses should be skilled in communication and language. As it is more important to develop a wide network of business, skills in a single language will help in growing in professional career and help in increasing product output.

A study of literature research work regarding AI for maintainable approach is replete with different aspects like overdependence on old data in ML processes, responses by different users if AI system gets disrupted, dangers of keeping huge processed data safely, adverse effects of AI, and complexities in computing outcome of interference plans.

These models are developed to suit to a particular environment and to consider the effect of rapid changing environment. To assess these models regarding their efficiency, capacity, level of usage, and many other aspects, the expertise of team members who have prepared DT systems are required as these models work in different ways as per their background of fields they are used in. These models are prepared for particular fields where they will be used and as per the need of that field. Hence, the requirement changes from field to field or application to application. Hence, a single model development and application will not give the required result. The various benefits of DT models in many fields, including easy availability, accuracy, and user friendliness, are the main reasons for their increasing popularity today. Earlier, these DT models where developed in such a way that they could be used by graphical interface; hence, it became more vibrant. Later, by combining many other tools and techniques, it became possible to use these models for a wider range.

If we compare these DT models with other developed models, then we can easily find that developed models use known information to find a solution. Whereas the DT models developed by redefining problems and applying the changing environment and reiterating the process to give a solution. The designers must work differently while developing these DT models as they are different from various other models.

In past decades, a lot of work was carried out for increasing its application base. With more and more applications used to model DT, there was a change in the essential rules, which is not healthy situation. DT model base is becoming wider, but it is missing out on innovations and further improvement in DT modeling technique. This is because the old developed methods are only used more and more rather than improving them by applying more improved methods.

In accordance with Brown (2009), while developing DT models, it is also necessary to add the effect of emotions on the results generated by DT models. Right now, DT models are working based on a principle that to process data, there should be the use of imagination, finding different new or old solutions and learning from experience. But while designing, DT models are developed in such way that the factor of effect of emotions is kept out while giving solutions. Also, it needs to consider the effect of insecurity in using DT models creating impact by their results. The impact of mental judgment on DT models also needs to be evaluated. Preferences or attitudes also should be resolved as they also impact on the design thinking process.

The involvement of design teams for the development of different models is increased. Different roles are assigned to different team members, which may affect DT modelling process. Also, social parameters have an impact on team members, causing changes in modelling and affecting individual team member performance, as well as the performance of the entire team. Effect of all these parameters need to be addressed (Dam and Siang, 2018).

4.6 SUMMARY

Artificial Intelligence (AI) and Machine Learning are the building blocks of tomorrow's technology. In today's world, AI has a variety of applications right from CRM to refining processes for different sectors. ML is the backbone of AI. The basic concept of ML is that we find an approach in such a way that the machine "learns" and can help us reach an optimized solution. Design thinking works on the basis of redefining a difficult problem, coming up with multiple solutions, and then selecting the most optimized one. Design thinking is a process in which a program is developed to process and update a massive amount of data every time before suggesting a solution to any given problem unlike human thinking where after a point of time we stop thinking and monotonously continue finding the solution to the problem rather than thinking.

4.7 CONCLUSION

AI is widely used in many sectors. Use of ML and high-speed performance of machines make AI a promising choice. As databases expand day by day, we need systems which can access, process the data at a high speed, and arrive at the solution of the problem immediately. We also need them to review and update the solutions as per new set of conditions and to help to reach to the optimized solution. Design thinking is one of the new developed ways and repetitive procedures. This procedure works toward keeping information regarding usage done by customers, experiment theories, and redefinition of difficult situation to detect substitute methods of strategies and solutions which cannot be noticeable at the time of initial level of understanding; and also it provides solution-based approach to the problems. To make machines learn and apply intelligently, we have to follow a procedure which mimics thinking. We humans remember a particular object by its noticeable characteristics like shape, color, weight, size, or any other special features. Our mind compares these things as soon as new object comes in front. Our mind first compares and takes a particular decision about an object. As soon as our mind sees a new object, it starts remembering the old object even though only few things matches with the new object. In this type of case, we block our thinking by replacing thoughts and observations of the new object. This is actually not correct or may lead us to misconception regarding new object and coming to a set of wrong decisions even if all features do not match. Hence, it is very important to check the object individually rather than just comparing with only similar or matching characteristics. It needs different approach of thinking to find solutions of a particular problem. Many instances occur in our life where our mind compares the past and new experiences and finds the necessary solution. What we need to understand is that instead of comparing each problem with the past experiences, we must come up with the optimized solution which would save us a lot of time and effort.

In this way, we can conclude that incorporating design thinking in AI and ML can help us arrive at the optimized solution always along with the vast amount of knowledge of this age. We need not write a fresh program for every sector as the machine has now learnt and applies its knowledge to solve the problem. Thus, we can see that design thinking in AI and ML is going to be prominent in the near future.

REFERENCES

Akerkar, R. 2019. *Artificial intelligence for business*. Springer.

Badke-Schaub, P., Roozenburg, N. and Cardoso, C., 2010, October. Design thinking: A paradigm on its way from dilution to meaninglessness. In *Proceedings of the 8th design thinking research symposium (DTRS8)* (pp. 39–49). DAB Documents.

Braha, D. and Maimon, O., 1997. The design process: properties, paradigms, and structure. *IEEE Transactions on Systems, Man, and Cybernetics-Part A: Systems and Humans*, 27(2), pp. 146–166.

Brown, T., 2009. *Change by design: How design thinking creates new alternatives for business and society.* Collins Business.

Dam, R. and Siang, T., 2018. *What is design thinking and why is it so popular*. Interaction Design Foundation.

Doepker, M., 2019, January. 5 ways students can graduate fully qualified for the fourth industrial revolution. In *World economic forum annual meeting*, 22–25 January 2019.

Dunne, D. and Martin, R., 2006. Design thinking and how it will change management education: An interview and discussion. *Academy of Management Learning & Education*, 5(4), pp. 512–523.

Dym, C. L., Agogino, A. M., Eris, O., Frey, D. D. and Leifer, L. J., 2005. Engineering design thinking, teaching, and learning. *Journal of Engineering Education*, 94(1), pp. 103–120.

Elsbach, K. D. and Stigliani, I., 2018. Design thinking and organizational culture: A review and framework for future research. *Journal of Management*, 44(6), pp. 2274–2306.

Evans, D. L., 1990. Design in engineering education: Past views of future directions. *Engineering Education*, 80(5), pp. 517–22.

Guzman, A. L. and Lewis, S. C., 2020. Artificial intelligence and communication: A human—machine communication research agenda. *New Media & Society*, 22(1), pp. 70–86.

Kietzmann, J. and Pitt, L. F., 2020. Artificial intelligence and machine learning: What managers need to know. *Business Horizons*, 63(2), pp. 131–133.

McGrew, J. B., 1970. *Future shock. Alvin Toffler*. Random House, 1970.505 pp. $7.95. The bulletin of the national association of secondary school principals, 54(349), pp. 123–129. (Toffler)

Norris, D. J. 2017. *Beginning artificial intelligence with the Raspberry Pi* (pp. 1–369). Apress.

Razzouk, R. and Shute, V., 2012. What is design thinking and why is it important? *Review of Educational Research*, 82(3), pp. 330–348.

Rifkin, J., 2014. *The zero marginal cost society: The internet of things, the collaborative commons, and the eclipse of capitalism*. St. Martin's Press.
Rizzo, A., Rossi, A., Lorusso, M. and Burresi, G., 2019. *More design thinking*.
Rotherham, A. J. and Willingham, D., 2009. To work, the 21st century skills movement will require keen attention to curriculum, teacher quality, and assessment. *Educational Leadership*, 9(1), pp. 15–20.
Simon, H. A., 1996. *The sciences of the artificial*. MIT Press.
Walch, M. and Karagiannis, D., 2020. Design thinking and knowledge engineering: A machine learning case. *International Journal of Machine Learning and Computing*, 10(6).

5 Learners' Learning Using Design Thinking Learning Approach (DTLA) with Machine Learning

D. Magdalene Delighta Angeline[1], *T. Prabakaran*[1] *and I. Felcia Jerlin*[2]
[1]Joginpally B.R. Engineering College, Hyderabad, India
[2]Holy Cross Engineering College, Thoothukudi, India

CONTENTS

DOI: 10.1201/9781003189923-7

5.1 INTRODUCTION

Technologies thrive on the needs of the learners as they are sprouting at a rapid speed. The educational sector and teaching-learning methods are affected and forced to adapt to the changing demands of the times. It is posing challenges to the facilitators to comprehend and design based on the changing needs of the times. Also, in the aftermath of the emergence of COVID-19, the educational sector is facing lots of challenges presented before it as the learners are returning to campuses. In order to address the increasing operational challenges, the educational institutions need to react, innovate, and renovate to meet the need of the learners. During the COVID-19 pandemic, the learners struggled openly on account of a lack of collective interaction with their academic peers, and this impacted the learners to reintegrate into the college life after pandemic.

As a result of COVID-19, the educational institutions faced challenges such as learner's knowledge issues, campus management, financial issues, and staff issues (Healy et al., 2020). The learners' knowledge requirements to be addressed related to the changes (Bovill et al., 2011). Researchers highlighted that serious deliberation must be given for transforming the manner of learning from face-to-face and reshaping the learners to adapt to the blended learning environment (Green et al., 2020). The facilitator completely has to amend their teaching environment, educational strategies, and class delivery, with ensuing, considerable outcome on the learners' learning experience (Hill & Fitzgerald, 2020).

The digital revolutionizing age forces us to confront the endeavors to improve their contributions as problematic changes in the public arena and new innovative developments sway the existing plans of action. Design thinking (DT) is a probable medium to comprehend the modernization problem before translating the resulting insights into practical solutions (N. Dragicevic et al., 2017). DT is a problem-solving approach that results in related solutions through ideation (T. Brown, 2008). The fundamental here is that the development is to be made for a problem after understanding the problem well and getting the correct solution that the people expect (S. Bell, 2008).

The introduction of design thinking in the teaching-learning method bestows more opportunities on learners to widen their skills to compete with the technologies confronting them. The core objective of DT is to offer learners an environment that can make them think and work like an expert designer and thus promote empathy, educational consciousness, and risk taking (Sharples et al., 2016). It is necessary to understand the importance of design thinking in achieving learning outcomes. For analyzing the educational needs of the learners during the course plan preparation, the significance of DT is hassled in the educational process. The facilitator's method of teaching and an organized way of lesson planning in making the learners learn a topic takes care of the concentrations and imaginative needs of learners.

In the education sector, the phrase "design thinking" was primarily connected with the manner designers think, but, in recent times, it has been employed by the other constructivist approaches that stress on reliable learning methods that are project-based, problem-based, and inquiry-based.

5.2 LITERATURE REVIEW

The development of many of these twenty-first-century skills is facilitated by teaching the learners with design thinking, and existing research supports this statement (Koh et al., 2015). Design thinking promotes high-quality theoretical understanding and creative thought (Barlex & Trebell, 2008; Gerber & Carroll, 2012; Schooler, 2004). Gerber and Carroll (2012) concluded that the act of fast prototyping strengthened the creativity beliefs within the design process. The design challenge facilitated higher-order thinking skills such as analysis, synthesis, and evaluation (Schooler, 2004). Moreover, many researchers put forward the notion that design thinking enhances meta cognitive skills such as reflection and self-regulation (Conlin et al., 2015; Kolodner et al., 2003; Sabag et al., 2014). Cochrane and Munn (2016) described design thinking process as a cycle: empathizing and observing, defining the problem, creating ideas, prototyping, and testing.

A study by Schiedgen et al. (2015) proved that 71% of respondents found design thinking improved working culture, and 69% found it made their innovation processes more efficient. Co-creation is a key component of design thinking, which turns into extremely efficient and is "enhanced" when used as part of the process (Hirano et al., 2013). The co-creation process of mental health supports and encourages learners and facilitators to use it for rebuilding the education system subsequent to the COVID-19 pandemic (Drissi et al., 2020).

According to Renard (2014), the term DT has its basis in diverse disciplines and is often, even though not entirely, linked with engineering, architecture, and associated design disciplines in premature literature spotlighted on DT. The quintessence of DT is to have users accustomed into the circumstances that make them imagine and work like an expert designer and thus promote public literacy, understanding, and educational awareness and risk enchanting (Sharples et al., 2016). According to Skaggs (2018), the tools' observation, experience, and question permit designers to recognize individual requirements and provide information to drive the formation of solution for the products and experiences that make human links through visuals, need-finding, or making sense. At the same time, as the consciousness of the designed experience rises, so does the yearning to make the procedure of DT available to a wider assortment of scenarios to analyze and determine any business or productivity challenge in a novel, shrewd, stimulating approach (Hodgkinson, 2013). Taheri et al. (2016) stated that, "differences amid experts concerning the common understanding of DT, let alone its anticipated learning results". The lack of theoretical clearness does not reduce speed of the espousal of DT in education.

Valentim et al. (2017) disputed that is vital to introduce DT in Computer Science and Software Engineering courses as a logical and imaginative procedure as it affords a human-centered view of technical artifact design. For itself, it lets facilitators to better train learners for the software development industry. Goldman et al. (2014) stated that in more than 60 US universities and colleges, DT education is imparted through training, workshops, and courses. Likewise, Callahan

(2019) examined that DT is being used in K-16+ curricula to promote the development of twenty-first-century skills, champed by the company IDEO and Hasso Plattner Institute of Design. DT has turned out to be a great experience in higher education owing to its prevalent consequences athwart many disciplines (Beligatamulla et al., 2019).

5.2.1 What Is Design Thinking?

Design thinking is making a difference and having an intentional process so as to get to new, related results that can produce positive impacts. Design thinking creates confidence in imaginative abilities and a process for transforming intricate challenges into opportunities for design. Gruber et al. (2015) defined design thinking as a "human centered approach to innovation that puts the observation and discovery of often extremely shaded, still inferred, human needs right at the vanguard of the innovation process".

Design thinking process undergoes two phases given in Figure 5.1: Problem phase and solution phase. The problem phase gives basic insights on how design problems are perceived, and the solution phase illustrates how to learn and deal imaginatively with pertinent knowledge within design-thinking-based problem solving. DT is defined as a human-centric process that builds upon the deep understanding of the end users on areas such as behaviors or tendencies to generate ideas or prototypes and to put out novel solutions to the world. The design thinking model revolves around the following steps:

- Empathize—Understanding the knowledge of the user for whom you are designing.
- Define—Unpacking the findings in the empathy step into needs, insights, and scope for a meaningful challenge.
- Ideate—Discovering a wide solution space
- Prototype—Developing a physical model
- Test—Evaluating the model and gathering feedback to refine the model

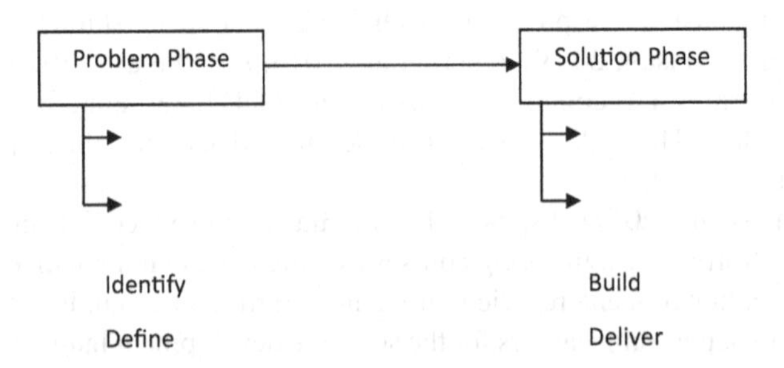

FIGURE 5.1 Design thinking process.

5.2.2 What Is Machine Learning (ML)?

Machine Learning is a core sub-area of Artificial Intelligence (AI), which offers a way for data analysis. With ML, computers discover perceptive information, devoid of being told where to stare. As an alternative, ML does this by influencing algorithms that learn from data in an iterative process.

Machine Learning is a method of data analysis which is a machine-centered approach that requires computational thinking. In Machine Learning, there are algorithms to make the machine to learn through data iteratively and make it able to find hidden insights from the data. By this, the machine can itself take decisions without being explicitly programmed in what to do in each case. The phases of Machine Learning are:

- Analyze
- Synthesize
- Ideate
- Tuning
- Validate

5.2.3 Design Thinking Versus Machine Learning

The stages of DT and Machine Learning are given in Table 5.1.

5.2.3.1 Step 1: Empathize and Analyze

The aim of the step of "Empathize and Analyze" is to understand the thoughts and experience of the users or the problem. In design thinking, it refers to understanding and capturing the user's sting point on a specific problem which helps to identify the degree at which it is probable with reference to the user. In Machine

TABLE 5.1
Stages of Design Thinking and Machine Learning

Design Thinking	Machine Learning
Empathize: Learn about the user or the problem	**Analyze:** Understand the basics and breakdown the needs into decisions
Define: Create POV based on the user needs	**Synthesize:** Combine separate elements so as to create a new whole
Ideate: Brainstorm to create as many as possible solutions	**Ideate:** Brainstorm to process the requirement into features and metrics
Prototype: Create a model of the solutions	**Tune:** Develop the model by fine tuning the model and boost the accuracy
Test: Share the prototype for collecting feedback	**Validate:** Reduce the overfitting in the model.

Learning, it refers to capturing the user's input verdicts and to discovering the variables and metrics that are useful for building better predictions. At this stage, the user's key decisions are identified and captured to find out which variables and metrics are needed.

5.2.3.2 Step 2: Define and Synthesize

In DT, this step refers to the formation of a Point of View (POV) based on user's or the problem's requirements and insights that principally refer to design, documentation, and authentication of the understanding of user's or problem's needs. In Machine Learning, it refers to the mixture of disconnected elements so as to generate a complete innovation. Here, data gathering, data exploration, and preparation are employed to better define the problems based on insights gotten from the data so as to prepare scalable solutions at a much smoother rate.

5.2.3.3 Step 3: Ideate and Envisage

In DT, it refers to thinking about as many solutions as is probable and then prioritizing those solutions on the basis of viability and customer values and implementation. In Machine Learning, with small sample data, various analytical models and algorithms are applied so as to observe which types of insights are present in the data. In ML, solutions are based on models that use data to generate predictions. The objective is to train the best performing model possible, using the processed data.

5.2.3.4 Step 4: Prototype and Tune

In DT, it entails providing the effective prototype to the user, involving the study of user's interactions with the creation, to see what works and where the users are facing problems and recognize what additional designs can be added in creation so as to improve the user experience. Machine Learning identifies where analytics insights are required and what additional data can be captured.

5.2.3.5 Step 5: Test and Validate

In DT, it refers to the monitoring the usage to resolve the efficiency of the created model or solution and to continuously improvise on the created model or solution, as feedback helps quickly to improve user experience. In Machine Learning, for the continuous improvement process from the user experience, over fitting is reduced. Here, as much data as possible is used to train, validate, and test the model.

5.3 DESIGN THINKING LEARNING APPROACH IN LEARNING ENVIRONMENT

DT is the process of discovering and solving problems with a human-centric approach putting human's needs, desires, and abilities at the center of the designing process. Design thinking is a prospect-focused learning, which means that it allows learners to transform their shared and environmental perspectives through design. The DT scaffold engages considering the world as a solution focused way and providing the self-confidence to resolve problems, which promotes critical thinking and imagination.

A DT approach spotlights on developing learners' imaginative self-confidence. Also, design thinking is used to improve the individual understanding. DT unites teamwork, system thinking, and a steadiness of imaginative and diagnostic practices. It assists learners to make the world a better place. Design thinking associates real-world problem-solving with learning environments. Facilitator and learners employ in hands-on design disputes that focus on:

- Developing understanding
- Endorsing action
- Encouraging ideation
- Extending metacognitive responsiveness
- Nurturing dynamic problem solving

During these processes, learners are persuaded to understand the problem, confront suppositions, and redefine problems. At the center of design thinking is the objective to breaking down the problem by analyzing and understanding how users interact with the problem and inspecting the circumstances in which they function. This involves asking questions and demanding suppositions. Once the learners have questioned and inspected the state of a problem, the solution-generation process will facilitate generating ideas that reveal the real limitations and aspects of that meticulous problem. Design thinking offers a way of excavating deeper. It facilitates spotlight research, prototyping, and testing problem and services to find new ways of improving the problem and design.

The facilitator acts as a guide in DT process whereby it focuses on learner-centered approach. The responsibility of the facilitator is to guide learners while they develop their own leadership and research abilities. The major purpose is for learners to learn how to learn and to become vigorous means of their own learning. The facilitator designs a set of rules for every phase of the learning and makes the learners to follow them. This whole process can be done online or offline, with the assistance of any constructive authoring tool or using any digital or paper tool. During the design thinking process, the facilitator persuades learners to see restraints as inspiration (Brown & Wyatt, 2010).

5.3.1 Challenges Faced by the Facilitator

A few challenges are there in specific to DT such as difficulty in understanding the concept, with a lack of capability and resources to initiate DT in the education system, and a lack of assistance in creating an appropriate task for learners. Some researchers also noted incapability to direct learners in the nonlinear DT process, predominantly the synthesis stage, and intricacy implementing design thinking in certain subjects and across curriculum. Some challenges are faced by the facilitator in terms of

- Design and development of learning experiences (curriculum)
- Learning environments
- Methods and tools
- System approaches

5.3.2 Design Thinking Integrated with Machine Learning

Design thinking is an imaginative act allowing facilitator to understand the act of creating an actually efficient learning environment. The facilitators need to be the entrepreneurial designers and re-designers of the traditional teaching method to transform the way education and learning are imparted to make them more pertinent, effectual, and enjoyable. The DT steps given in Figure 5.2 are as follows:

- Experience—Understanding about a given topic/problem
- Envisage—Predicting and defining ideas for the topic/problem
- Create—Ideating a solution
- Evaluate—Building a prototype
- Share—Integrating the model

5.3.2.1 Experience

The first step in design thinking is experience that is the understanding or knowledge about a given topic or problem. The learners collect all information related to the topic or problem, make their research and experiments, and start the teamwork with their group. Here, the learners learn the topic in an intellectual, collective, and emotional way, developing empathy, creativity, and a shared viewpoint of the topic in their group. This step helps learners to intensely attach with their group, find out how a given problem subsists in their group, and how their group relates to it. It permits learners to recognize the main traits of their group and of the problem to be considered while discussing possible solutions.

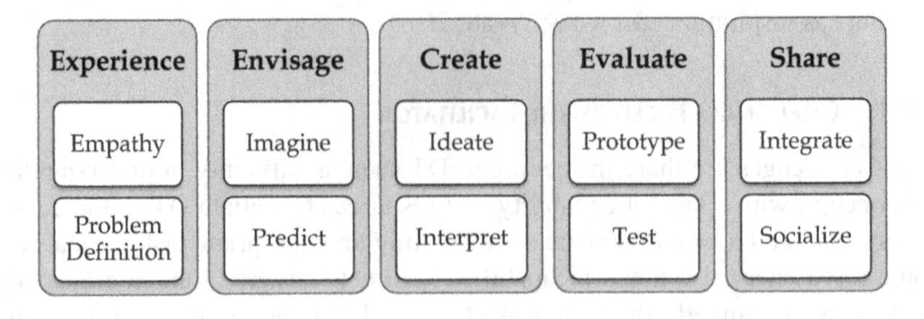

FIGURE 5.2 Steps in design thinking.

This phase is robustly linked to the Inquiry-Based Learning method, even though the whole DT process is in sequence with the similar inquiry standards. For every topic, different situations will be offered in this stage. This activity increases the learner's inquisitiveness on the topic allowing them to be motivated and stay connected to the topic, eager to learn more. To help the learners understand much more about a topic, the facilitator shows videos, images, 3D Interdisciplinary map of science ideas, concept maps, virtual classroom walls, animations and/or simulations, etc. The facilitator can also develop a story or an event related to the topic that widens learners' memories or experience. The presentation about a topic is another way to get familiarized with the topic and to extract the variables related to the topic. The learners are allowed to go out of the classroom environment and visit real people who experience the topic at hands every day, listen to what they have to say, ask them how they think the problem could be solved, and then start creating solutions accordingly. Later, the learners are encouraged to register all their ideas and the conversations with their group.

5.3.2.2 Envisage

The envisage phase widens the thinking ability and induces the imagination of the learners on a problem. Each learner has different view and ideas over a problem. Brainstorming sessions and mind mapping activities explore their possible imaginations and find solutions to challenges and then aid them in identifying the best solution from a pool of ideas. Discussion on the topic facilitates the learners to seek answers recalling previous knowledge and to share the knowledge with the groups, increasing the capacity of retaining new information. The concept of the topic is connected with what the learners know, and a concept map is created.

5.3.2.3 Create

After understanding the problem or a topic, the ideas of all learners are discussed among the peer group and defined. This phase helps to put together all the information gained during the experience step and scrutinize the data to produce a meaningful statement. This is the most important phase where the learners' ideas are collected together and categorized based on the feasibility and time constraints of the ideas. The ideation phase helps learners to discover a wide solution space from a broad diversity of ideas defined by the learners.

To evaluate the problem in Machine Learning, further, deep understanding of the problem is to be made through which the variables and metrics are identified that are the better predictors of the problem. The entire group can share their ideas and understanding about the topic to identify the variables that impacted the problem well and frame it to take a decision with Machine Learning algorithms.

5.3.2.4 Evaluate

In DT, evaluate refers to scrutinizing the usage to resolve the efficacy of the solution. In this phase, a prototype is built with which a user can interact. Ideally, one is biased toward something a user can experience. From the huge repository of ideas, prototypes are built for testing. For putting this into practice, Machine

Learning model is employed. The feedback system is implemented to improve users' experience to continuously make improvements to the solution. With Machine Learning, overfitting in the model is avoided which facilitates the continuous improvement process from the user experience and analytic model tuning perspectives.

5.3.2.5 Share

Sharing is the final phase where the learners share their solution to a problem with others and spread their outputs to groups of people. Here, the learners teach others about what they have learned to bring awareness through their designs. Learners have been exposed to a new way of learning that has given arms to their imagination and brainwave, and they have emerged from their comfort zones. Through sharing their work, both learners and facilitators can ascertain an intellect of achievement and satisfaction in their work and of themselves.

5.3.3 Methodology

Machine Learning facilitates a machine to learn from data without human intervention, develops performance from a practice, and foresees things without being definitely programmed. ML enables a more incorporated, well-versed design process when developer employs with data. For a realistic solution to be created, it's indispensable to resolve the problem immediately by creating the right structure around problem and capturing it. After the question is captured absolutely, the problem can be addressed and analyzed to generate case studies around the problem areas and envisage an AI-enabled solution. When giving training data to ML algorithm, a mathematical model is generated, and prediction and decision are made by the machine devoid of being openly programmed. Also, during training data, the more machine will work with it the more it will get experience, and the more it will get experience the more efficient result is produced.

Initially, the problem is understood, and use cases are studied. After understanding the problem, variables are formulated, and data is collected relevant to the problem. Further, questions are raised to understand and identify the Machine Learning algorithm that suits the problem to devise a solution for the problem. The ML algorithms researchers have been developing for decades, when applied to today's web-scale data sets, can yield amazingly good forms of intelligence.

5.3.3.1 Procedure of ML

- Collect the data related to the problem
- Discover and decide the most appropriate ML algorithm for the problem
- Divide the data into subsets
- Preprocess the data to clean the data
- Enhance the data
- Train the data
- Test the model and save it for future use

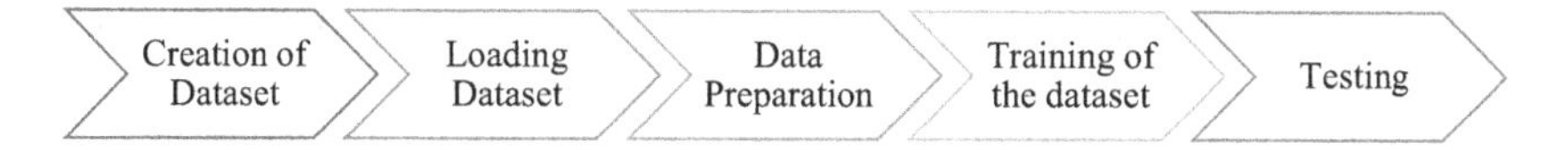

FIGURE 5.3 Steps of Machine Learning process in Envisage phase.

Figure 5.3 explains the steps carried out with the Machine Learning model under the create phase. From the ideas collected on a problem, the variables are listed out to formulate a dataset. After creating a dataset, the dataset is loaded into the Machine Learning model for analysis. The dataset undergoes data preparation stage in which the preprocessing of data is performed. Here, the data cleaning is done, and relationships of the data get well-defined. Then the data is further divided into 80% and 20% for the training and testing process using train-test split technique.

The train-test split is a technique for estimating the performance of a Machine Learning algorithm. The method entails taking a dataset and dividing it into two subsets. The first subset is used to fit the model and is referred to as the training dataset. The second subset is not used to train the model; instead, the input element of the dataset is provided to the model, then predictions are made and compared to the expected values. This second dataset is referred to as the test dataset.

- **Train Dataset**: Used to fit the Machine Learning model.
- **Test Dataset**: Used to evaluate the fit Machine Learning model.

5.4 CONCLUSION

Design thinking is the most advantageous approach in the education sector facilitating the learners to learn with utmost interest and helps learners in understanding deeply about a topic or problem in a well manner. Design thinking is equally a practice and way of thinking that developed from investigations on "design thinking" into a problem-solving approach principally adopted in industry to a prevalent way of dealing with depraved problems. This design thinking is to be employed in re-creating an emergent role in the education sector. Inspirations for utilizing DT in education are classically complicated. Educators are anticipating unforeseen ideas, chic solutions, and new notions to make a learning or development experience easy in a new and energizing design and persuade convenient dexterities and competencies among the learners.

The facilitators utilize human-centered design techniques integrated with DT approach in a structured way to understand their learners better, making learners feel more connected and widen their ideas in the changing learning environment. The facilitators identified that they had revolutionized their techniques of teaching, mostly using design thinking as a technique of inquiry in the learning environment. They monitored transforms in the way they plan, teach, and evaluate owing to the new skills they acquired. Also, they were more positive and keen to step out of their comfort sector to direct and work together with their coworkers. The five phases interlinked with Machine Learning assist to find the way of development from recognizing a design to verdict and build a solution. It is an intensely human approach that relies on individual ability to be perceptive, to construe what is observed, and to build up ideas that are expressively significant to those who designing for. Among learners, the most noteworthy impact monitored right through the assessment was that of enhancements in meta cognition acquaintance and expertise. When learners engrossed themselves in the DT process, they spent more time reflecting on their learning and applying their knowledge in an assortment of scenarios. DT is a resourceful approach for organizing contradictory ideas, recognizing remarkable needs and general purposes, making creative exploit of varied backdrops, enhancing understanding, and developing a collective visualization in a learning environment.

REFERENCES

Barlex, D. M., & Trebell, D. (2008). Design-without-make: Challenging the conventional approach to teaching and learning in a design and technology classroom. *International Journal of Technology and Design Education*, 18(2), 119–138.

Beligatamulla, G., Rieger, J., Franz, J., & Strickfaden, M. (2019). Making pedagogic sense of design thinking in the higher education context. *Open Education Studies*, 1(1), 91–105.

Bell, S. (2008). Design Thinking. *American Libraries*, 39(1&2), 44–49.

Bovill, C., Cook-Sather, A., & Felton, P. (2011). Students as co-creators of teaching approaches, course design, and curricula: Implications for academic developers. *International Journal for Academic Development*, 16(2), 133–145.

Brown, T. (2008). Design thinking. *Harvard Business Review*, 86(6), 84–92.

Brown, T., & Wyatt, J. (2010). Design thinking for social innovation. *Development Outreach*, 12(1), 29–43.

Callahan, K. C. (2019). Design thinking in curricula. In *The International Encyclopedia of Art and Design Education* (pp. 1–6). John Wiley & Sons. American Cancer Society.

Cochrane, T., & Munn, J. (2016). EDR and design thinking: Enabling creative pedagogies. In *Proceedings of EdMedia 2016-World Conference on Educational Media and Technology* (pp. 315–324). Vancouver, BC, Canada: Association for the Advancement of Computing in Education (AACE). Retrieved April 3, 2018, from www.learntechlib.org/p/172969/.

Conlin, L. D., Chin, D. B., Blair, K. P., Cutumisu, M., & Schwartz, D. L. (2015). Guardian angels of our better nature: Finding evidence of the benefits of design thinking. In 122nd ASEE Annual Conference & Exposition. Seattle, WA.

Dragicevic, N., Lee, W. B., & Tsui, E. (2017). Supporting service design with storyboards and diagrammatic models. Proceedings Theory and Applications in the Knowledge Economy Conference, pp. 457–469.

Drissi, N., Ouhbi, S., Idrissi, M. A. J., & Ghogho, M. (2020). An analysis on self-management and treatment-related functionality and characteristics of highly rated anxiety apps. *International Journal of Medical Informatics*, 104243.

Gerber, E., & Carroll, M. (2012). The psychological experience of prototyping. *Design Studies*, 33(1), 64–84.

Goldman, S., Kabayadondo, Z., Royalty, A., Carroll, M. P., & Roth, B. (2014). Student teams in search of design thinking. In *Design Thinking Research* (pp. 11–34). Switzerland: Springer International Publishing.

Green, J. K., Burrow, M. S., & Carvalho, L. (2020). Designing for transition: Supporting teachers and students cope with emergency remote education. *Postdigital Science and Education*, 1–17.

Gruber, M., de Leon, N., George, G., & Thompson, P. (2015). Managing by design. *Academy of Management Journal*, 58(1), 1–7.

Healy, E., Kinsella, D., & Cremin, M. (2020). *Understanding the Impact of Covid-19 on Higher Education Institutions*. Dublin: Deloitte. Retrieved September 30, 2020, from https://www2.deloitte.com/ie/en/pages/covid-19/articles/covid-19-on-higher-education.html.

Hill, K., & Fitzgerald, R. (2020). Student perspectives of the impact of COVID-19 on learning. *All Ireland Journal of Higher Education*, 12(2).

Hirano, T., Ishizuka, A., & Sakaguchi, K. (2013). Innovation activities by co-creation process. *FUJITSU Scientific & Technical Journal*, 49(4), 391–396.

Hodgkinson, G. (2013). Teaching design thinking. In J. Herrington, A. Couros, & V. Irvine (Eds.), *Proceedings of EdMedia 2013-World Conference on Educational Media and Technology* (pp. 1520–1524). Victoria, Canada: Association for the Advancement of Computing in Education (AACE).

Koh, J. H. L., Chai, C. S., Wong, B., & Hong, H.-Y. (2015). *Design Thinking for Education*. Singapore: Springer Science+Business Media.

Kolodner, J. L., Gray, J. T., & Fasse, B. B. (2003). Promoting transfer through case-based reasoning: Rituals and practices in learning by design classrooms. *Cognitive Science Quarterly*, 3(2), 119–170.

Renard, H. (2014). Cultivating design thinking in students through material inquiry. *International Journal of Teaching and Learning in Higher Education*, 26(3), 414–424.

Sabag, N., Trotskovsky, E., & Waks, S. (2014). Engineering design projects as a reflection promoter. *European Journal of Engineering Education*, 39(3), 309–324.

Schiedgen, J., Rhinow, H., & Köppen, E. (2015). Without a whole–The current state of design thinking practice in organizations. Study Report, Hasso-Plattner-Institut Potsdam.

Schooler, S. R. (2004). A "chilling" project integrating mathematics, science, and technology. *Mathematics Teaching in the Middle School*, 10(3), 116–121.

Sharples, M., de Roock, R., Ferguson, R., Gaved, M., Herodotou, C., Koh, E., . . . Wong, L. H. (2016). *Innovating Pedagogy 2016: Open University Innovation Report 5*. Milton Keynes: The Open University

Skaggs, P. (2018). Design thinking: Empathy through observation, experience, and inquiry. In E. Langran & J. Borup (Eds.), *Proceedings of Society for Information Technology & Teacher Education International Conference* (pp. 1168–1172). Washington, DC: Association for the Advancement of Computing in Education (AACE), 2008.

Taheri, M., Unterholzer, T., Hölzle, K., & Meinel, C. (2016). An educational perspective on design thinking learning outcomes. In *ISPIM Innovation Symposium* (p. 1). The International Society for Professional Innovation Management (ISPIM). Innovation Forum, Boston, MA.

Valentim, N. M. C., Silva, W., & Conte, T. (2017). The students' perspectives on applying design thinking for the design of mobile applications. In *Proceedings of the 39th International Conference on Software Engineering: Software Engineering and Education Track* (pp. 77–86). IEEEXplore, IEEE Press, Buenos Aires, Argentina.

6 Design Thinking to Mobile App Development

Anupama Namburu and Prabha Selvaraj
School of Computer Science Engineering,
VIT-AP University, Guntur, India

CONTENTS

6.1 INTRODUCTION

Institutions devoted to building innovative and ingenious applications should consider design thinking as a solution. Design thinking creates an analytical and imaginative thought process to educate individuals about the technology design environment effectively. Design thinking is regarded primarily as a way towards

DOI: 10.1201/9781003189923-8

comprehending challenges but instead developing creative, persuasive ideas [1]. Design thinking explores fundamental underlying issues but synthesizes responses that use a systematic sapient methodology that includes imagination, brainstorming, and execution.

Recognizing and addressing the overall demands of people and possessing one distinct style to reason with creativity are the anticipated primary consequences of design thinking. Educating learners on the value of revision would aid with the overall development, building capabilities that may enable learners to become innovators. Design thinking encourages the overall development of communication competencies. Therefore, as a result, it promotes brain innovation. However, due to their unique structures available for performing operations for distinct style of creativity, understanding design thinking becomes extremely difficult. Because given those challenges within DT programs, teachers now attempt primarily to educate solely based on the fundamental principles and distinct methodologies of Creative Thinking, followed by practical tasks that utilize those principles learnt [2].

As design thinking has a complicated nature, it becomes very complex. To teach the learners about DT, the instructors must innovate on various methods to overcome the complexities of DT. One way could be leading the learners to different concepts, components, and processes of DT and then moving on to practical activities to apply the concepts learned earlier. Since the advent of technology in recent years, it has become evident to focus on design thinking in mobile applications.

Business Competitiveness has been increasing exponentially due to various design products and services. There has been a motion of ideating an application that meets the needs of consumers rather than making previously established philosophies more attractive. Design thinking came as a support for problem-solving that applies imaginative perspectives to a diversity of grounds to benefit industries and personalities taking innovation to its peak. DT provides the inventor's susceptibility and methods to agree with what is scientifically achievable with customers' wants [3]. Furthermore, this professional approach can be transformed into client value and new market opportunities.

DT comprises numerous participants in the procedure, besides bargains and trials' philosophies. It encourages iterative idea cohorts, ongoing inquiry, and regular broadcasting, with the development team delivering their results towards the regions. As a result, design thinking seems to be a powerful tool for including multiple participants, especially technology builders, for meeting customers' demands.

Design thinking could be classified into three broad approaches rather than being broken into sequential steps. Inspirations, conceptualization, and deployment are the three iterative phases of design thinking [4]. The initial phase is Inspirations, which describes issues that emerge from encounters between clients and drive the quest towards ideas. Ideation is the subsequent step in producing, implementing, and evaluating a range of solutions-oriented concepts. At the end, design thinking's penultimate deployment step lays forth a strategy for developing and implementing the chosen notion.

In terms of implementation, DT aids in the development of novel capabilities. This, in turn, encourages the individual to step outside of their comfortable region and consider different ideas and design additional functionality and layouts for the app [5]. Whenever contrasted to alternative technology design approaches, this demonstrates the benefit of DT. Design thinking is a process based on a human-centered approach that closely monitors and identifies individual demands to improve their capacities [6]. As a result, the research design is a fundamental technique for mobile app creation to provide learners with the essential competencies to work in the application business.

Many authors have successfully applied design thinking for app development. The disaster management using mobile application is discussed by Suzianti et al. [7]. Design thinking used for designing course content and method of evaluation to improve students' performance is studied by Hsu et al. [8]. The author designed a charity application for collecting donation [9]. Schiele et al. discussed how design thinking is used for educating students in marketing [10]. An easy-to-use attendance system for university was developed [11]. A thematic study of information was needed for self-management of women with breast cancer [12]. A mobile application using design thinking was used to identify patients' issues in palliative care [13]. Design thinking for mobile hospital using Lean and Six Sigma for making patients more satisfied was explained [14]. Slam et al. has reviewed different mobile apps developed for COVID-19 pandemic using design thinking [15]. The author explains the usage of design thinking for developing project management by developing a mobile application [16].

Researchers widely acknowledge the necessity for pupils to acquire critical thinking abilities. You believe that making instructors aware of student deficiencies and modifying how the evaluation of the student takes place in learning are vital for enhancing students' critical thinking abilities.

The steps in design thinking to app development include the following:

A. Ideate and experiment

Based on the deep user understanding achieved during the empathy stage, cross-functional teams can quickly ideate and iterate on new app designs, going wide in pursuit of more innovative and effective responses to users' concerns. Using a number of reusable UI elements, the Atlas UI Framework allows you to quickly construct and update screens. By seeing programs live while they're being constructed, developers can meet with end users, get feedback, and make quick changes.

B. Keep user needs front and centre

Overall focus on users throughout the design and development process is necessary for efficient apps that meet both user and business goals [17]

With an integrated collaboration platform, business users can actively participate throughout the application lifetime to begin gathering feedback and iterating and share working prototypes or MVPs with users immediately away.

A built-in feedback loop in the development environment allows you to collect data on a continuous basis and assist in making the transition from prototype to production as smoothly as possible. Inbuilt tools that provide solutions are preferred to manage the user requirements and develop profitable programs by using high availability and scalability straight out of the box.

C. Promote consistency and reuse

UI team needs to create an enterprise-design language that developers can use in their apps, enhancing consistency and reuse across the board. Create a corporate design language with a customized set of page templates, building blocks and widgets, as well as a set of brand guidelines. Using reusable UI elements based on best practices, empower developers without UI design expertise to offer exceptional UX.

6.2 CUSTOMER NEEDS OF APPS

Identifying your target market is the first step in conducting a purchaser survey. Your market research should help you understand your clients better. More customer research can help you establish a more specific image of them and figure out how to reach out to them. It will also highlight important factors for your client, such as proportion, gender, age, occupation, income, residence, and leisure sports. Once you've figured out who your customers are, you can figure out what motivates them to buy products and services. Consider whether their decisions are influenced by the desire to buy, personal family needs, financial limits, social or emotional needs, or brand preferences. You may also need to know how they store in addition to why they store. To learn about your clients' preferred method and way of purchasing, keep in mind whether they save online, over the phone, or in stores, and if they purchase spontaneously or cautiously.

Find out what your clients' financial capabilities and spending patterns [18] are. Consider their average earnings [19] and the percentage of their income that they spend on the services or products you sell if they budget. Learn about your clients' perceptions and expectations of your business and competition by asking them what they think about you and what your customer service competitors' services and products are [20]. Patron cravings are described as the motivating factors that lead people to purchase your product or service. It is critical to comprehend the causes behind a customer's selection before deciding on their desires. It is vital to recognize who your clientele is to better understand their needs. A better understanding of the target audience can be accomplished by defining spending, earning patterns, and customer desires and segregating them based on the industry needs or the characteristics.

The more you know about your customers, the easier it is to develop your brand positioning [21] around their needs and to benefit your business in the following ways: give speedier responses—one of the most typical requests from customers is for real-time assistance in determining the problem, and by figuring out the desires of your customers, you may provide a faster and effective assist.

Improve your products and services—customer research can help you understand why people shop the way they do. You can research the areas where you're missing out and develop a compelling USP [22]. The information can be utilized to enhance products or services in order to meet the needs of customers. Reduce the frequency of assistance tickets—designing products and services around the needs of target clientele ensures effective solutions to client problems. A way to identify client needs is spotting patron needs, which includes a deep research across your industry and asking your clients plenty of specific questions.

Analyzing customer psychology [23] may help your company provide better customer service, develop long-term relationships, and maintain a steady stream of purchases. The most crucial way to anticipate is to do a thorough examination of clients' aspirations and needs. The buyer needs' analysis is a comprehensive examination that can assist your business in determining what cost your customers require from your services or products. It presents treasured insights of your target audience, which can be inculcated within the logo positioning to ensure providing an amazing purchaser price. Here are four simple measures to follow in order to successfully meet client needs. Surveys, interviews, attention corporations, and social listening can all be used to determine what customers want.

Distribute—Once you've identified the wishes, you can distribute them to the appropriate groups and departments.

Create—Tailor product capabilities and specific content to meet the wants of patrons. Obtain—get client feedback on a regular basis to see how your efforts are meeting their expectations. Client fulfilment is a critical component of achieving customer requirements. Learn more about how to create a successful customer experience strategy to achieve your goals. To align with core groups, corporations are making strenuous efforts to identify consumer desires and accommodate them as early as feasible. Seventy-six percent of customers expect businesses to understand their needs. With business working on a cyclical structure of anticipating and meeting the supplying needs, has tremendous impact on the market.

The effectiveness of a consumer understanding based on their desire is determined by two variables. To begin, construct consumer personas and determine what purchaser inputs are required to develop forward-thinking products. The second step is to determine how to acquire consumer inputs and comments. Customer research can be done in a variety of ways to identify the factors that influence purchasing decisions, such as consumer interviews that are the simplest approach to collect customer feedback. You could have interaction directly with clients who're using your product or who've selected to shop for it.

Companies' milestones are taken into consideration as they are the defining factors for obtaining inputs from the customers.

Cognizance businesses—recognition agencies are made up of a small group of people whose main focus is on a specific product or issue. Businesses place a greater emphasis on qualitative or quantitative surveys as they provide more reviews and motives.

Surveys—Survey research helps firms gain a better understanding of their market position in terms of meeting the needs of their target customers. Client desires come in a variety of shapes and sizes, and organizations must be aware of them in order to compete in today's market. Client demands, in general, should be analyzed for providing them a better experience by exceeding their expectations. When you anticipate your clients' needs, you can generate content, enhance your product features, or expand your offers to meet their needs sooner.

Customer requirements [23]—Customer demands can be classified based on the demographics of the marketplace's clients. Client requirements, on the other hand, can be divided into two categories.

Product requirements are related to and centred on the product. If your product meets your customers' needs, they become potential customers, and vice versa. Customers often set their budgets for any product purchase, thus price is one of the most important aspects of product wishes. Clients look for features that can solve their problems and ensure that the product continues to work reliably even after they have stopped using it.

Effectiveness—The product must be capable of streamlining the process to maintain a track of time. Companies' desires confer with the customers' emotional needs, and this have a direct impact on their purchase patterns on the customer.

6.3 APP DESIGN SPECIFICATIONS

An app specification must always be clear and easily understandable for anyone. Making separate headings for separate topics, mentioning details about the company and the app, and the goals and targeted audience are really important to specify. Let's take a detailed demonstration of that here. Here, you need to mention a few details about the company such as the nature of business, the goal of the company, and years of experience, and it would also be really helpful if you can mention any of the previous successful projects in app design.

A. App Concept/Research Context

The first step is to mention the main concept or idea of the project. You can provide some statistical information regarding the usage or need for the app to specify why this could be really important and useful thus successful. If it's an app which is going to be created as a competitor to a prior existing app, then specify the bonus features and reasons why this app could be better than the existing app.

B. Benefits, Convenience, and Privacy

Here, you specify the advantages and uses along with mentioning how the app will be beneficial to the users. Then you need to explain the app's convenience design [24] so that users of all generations can use the app with ease.

A detailed explanation of how the privacy design affects [25–28] the usage of the app needs to be given. The privacy risks involved must also be mentioned.

C. Target Audience

Mentioning the demographic that the app is targeting is really important, which is what you will be doing in this section. Along with that you also need to mention in which platforms or technical devices these "Audience" will be able to access the app such as android and iOS.

D. Model Specification/Estimation

By analyzing the data obtained through research, identify the problem and modify the app to increase acceptance. Now based on this, estimate the effects of benefit appeals, privacy design, and convenience design [29].

E. Display of Features and UI

Showing the visual appearance of the features and UI in the app as a template would be vital tool. Because visual description gives a better appeal to the document, this will also be helpful for the developers to implement.

F. Budget and Timeline

Even though estimating the time and cost is a difficult task, you should at least provide approximate estimations. Like for example if the app might take 3–4 months to build and you are not sure even if that is sufficient time, you can just mention that the project will be implemented under 5 months. The same goes with the cost, you can give the closest estimation.

One should mention how the acceptance criteria conditions [30] have been met and then identify the people in charge of the project.

You should try to mention as many details as possible about the project, but, most importantly, you should not forget to mention the important features like target audience, market values, and usefulness of the app.

6.4 APPLIED CREATIVITY FOR APP DESIGNING

The most commonly used mobile operating systems are Android (71.6%), which was developed by Google, and iOS (29.5%), which was developed by Apple as per the statics of Stat Counter worldwide. Whenever you want to create an application, User Interface (UI) and User Experience (UX) are mainly important, and they should be the primary factors. Because the complete app review mainly depends on UI/UX, to design the UI of an application first you need to select app category in which it comes under. For example, if the app to be designed is an Ecommerce app, then in the initial phase you need to do some research on the already existing apps which are popular and have rich UI. The most used and trusted E-Commerce platforms are Amazon and Flipkart. First thing you need to do is go through the app, move along all the screens of the app, and see how the components in the screen are displayed in a particular manner and which are user friendly.

After reviewing the existing apps if you are not satisfied with the UI in those, then go to dribble website which has plenty of mockups and UI designs of both web and mobile apps. After reviewing, you must think about our own design and make some wireframes of our design and then present it in the Graphic designing software such as Adobe-XD, Figma, and Sketch. You should use this software to custom design our apps. They will give the best User Interface as well as User Experience for our apps if you utilize them in an effective way. The most important thing to get a good design for our app is to place all the icons and text in an order and make sure that they are not going out of the screen while running the app. There are plenty of no-code apps where you can create apps without coding. Also if you want to design the application with code, then you can use native app development or cross-platform app development, and it depends on us.

If you're going to create that app design using native app development, then you must learn either java or Kotlin to design your app only for Android (or) swift for iOS app development. Otherwise, if you want to create a cross-platform app which runs on both Android and iOS platforms you must learn React-Native, Flutter, Ionic, etc. The app design must be more creative, ensuring that it doesn't look like any other existing apps, only then your app will be unique with good looking UI. To create web app, you should have coding experience in HTML, CSS, JS and also having some knowledge in JS Frameworks is good because designing a web app using only HTML, CSS, and JS is not good, and UI will be not better. If you use any JS library or framework to design a web app, then you can use third-party libraries in it for the icons, and for additional usage in our app, you can also use CSS Frameworks such as Bootstrap and Tailwind for easily styling our app with awesome design and cool animations with simple code.

Creativity is the key factor while you design an application because it will reach to the users when the app design is excellent, and then your app will be used by many users. The Wallas model is one that provides an explanation of the creative process (1926). The four-stage model of creativity proposed by Wallas is regarded as a foundational work [29]. The four processes are: 1) preparation, 2) incubation, 3) illumination, and 4) verification [29–30].

1) In the preparation stage, data and knowledge are acquired to gain a better understanding of the subject.
2) The information gathered in the preceding step is processed both consciously and unconsciously in the incubation stage.
3) During the illumination stage, the creative concept becomes obvious.
4) At the end, in the verification stage, the creative idea is validated and used.

6.4.1 Fostering Creativity

Figure 6.1 shows that Boden [31] made another distinction in creativity. There are three types of creativity, which are Combinational, Exploratory, and Transformational.

Creativity Type	Way of Fostering
Combinational	Combinations of existing designs getting more insight Useful for evaluation
Exploratory	Familiarize with a specific style Apply the same in a different way
Transformational	Familiarize with a particular style Familiar with Transformational creativity Understanding how it works Share the design Self confidence

FIGURE 6.1 Types of creativity (based on Boden's view 2009).

6.5 PROTOTYPING OF APPS

Prototypes are a replica of a final finished product, typically without code. It gives us an idea of how the end product ought to be. Prototyping is an important exercise that brings a sense of how the app is going to work by showing user flows, layout, and the design of the application. It's an essential part of developing an application as it enables you and your group to explore ideas and share comments at the early stage of building an application.

6.5.1 Types of Prototyping

Low-fidelity prototyping: Low-fidelity prototypes are often in paper form and do not permit interactions between users.

Paper Prototyping is a quick and easy way to illustrate a concept of your product using just a pen and paper. This is also the most cost-effective method of prototyping. With paper prototyping, you can sketch the primary layout of the screen, with basic features and shapes.

High-fidelity prototyping: High-fidelity prototyping is computer-based, and the interaction between the user and the application is feasible. High-fidelity prototypes bring you one step closer to a real representation of the user interface. It is assumed that high-fidelity prototypes are far more efficient at gathering real human performance data. They are also preferred for demonstrating the application to the clients.

6.5.2 Essential Tools for Prototyping

Here are some useful tools for making a prototype [32–33].

Figma: Figma integrates everyone involved in the design process, allowing teams to produce better products in less time. Figure 6.2 represent the logo of the Figma tool [34].

FIGURE 6.2 Figma (Source: Figma 2015).

FIGURE 6.3 Invision (source: invisionapp 2011).

Invision: InVision Freehand is the transformative visual collaboration tool disguised as an online whiteboard. Figure 6.3 represents the snapshot of the Invision tool [35].

Adobe XD: Adobe XD is a vector-based user experience design tool created and published by Adobe Inc. for online and mobile apps. Figure 6.4 depicts a snapshot of the Adobe XD for mobile app tool [36].

Axure: Axure is a dedicated rapid prototyping tool that enables any person with an elementary knowledge of software to create simple wireframes [37]. Figure 6.5 represents Microsoft Azure.

Sketch: Sketch provides a collaborative design process from the basic pixel to picture perfect design allowing rapid prototyping and developer hand-off [38].

FIGURE 6.4 Adobe XD (source: Adope 2016).

FIGURE 6.5 Microsoft Azure (source: azure.microsoft 2014).

6.5.3 Steps for Prototyping

Prototyping steps are shown in the Figure 6.6.

Step 1: Requirements of App
A prototyping model begins by analyzing the requirements. In this phase, the system requirements are defined in detail. Throughout the process, system users are questioned about what they expect from the system.

Step 2: Quick App Design
The second stage of prototyping is a preliminary design or a quick design. In this phase, a simple design of the system is made. It's not a complete design, and gives a brief idea to the user on how the system works.

Step 3: Building an App Prototype
In this phase, a small working model, which is called a prototype, is built. The prototype is designed on the basis of information gathered from quick design.

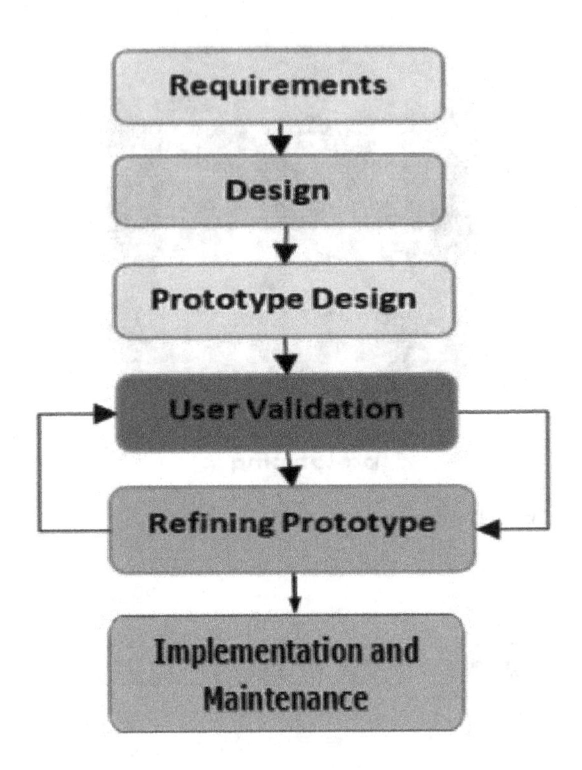

FIGURE 6.6 Different phases of prototyping model.

Step 4: Initial User Validation of App

In this stage, the prototype is introduced to the client for an initial evaluation. It helps to figure out the strength and weaknesses of the prototype. Suggestions and areas of improvement are collected from the client and submitted to the developer team.

Step 5: Refining Prototype of App

If the user is dissatisfied with the current working model, you need to refine the prototype based on the feedback and suggestions from the user.

This phase is not completed until all the requirements stated by the user are satisfied. Once all the refinement is done and the user is satisfied, a final system is developed based on the prototype that is approved.

Step 6: Implementation and Maintenance of App

Once the final system is developed based on the final prototype, it is fully tested and rolled out to production.

6.6 APP ARCHITECTURES

Application architecture is a collection of tools and methodologies for designing a well-structured mobile application that meet the needs of the industry and the user. The mobile apps are used on varied device configurations; hence, it is important to keep in mind all the consideration when designing the architecture of your app.

6.6.1 Layered Architecture

The architecture of mobile application is structured in different layers, as shown in Figure 6.7.

Presentation layer:

The presentation layer concentrates on the UI components and the interfaces that process the components. The presentation layer plays a vital role in the success of the app design, as the end user gets to visualize the app. While thinking about the design with respect to presentation layer, one should focus on the architecture, suitable data format and use effective data validation techniques to protect your apps from inputting inaccurate data.

Business layer:

The flow of the app design, the business units and the app building block are present in the business layer. App design usually follows MVC architecture where the business logic is separated from presentation. The

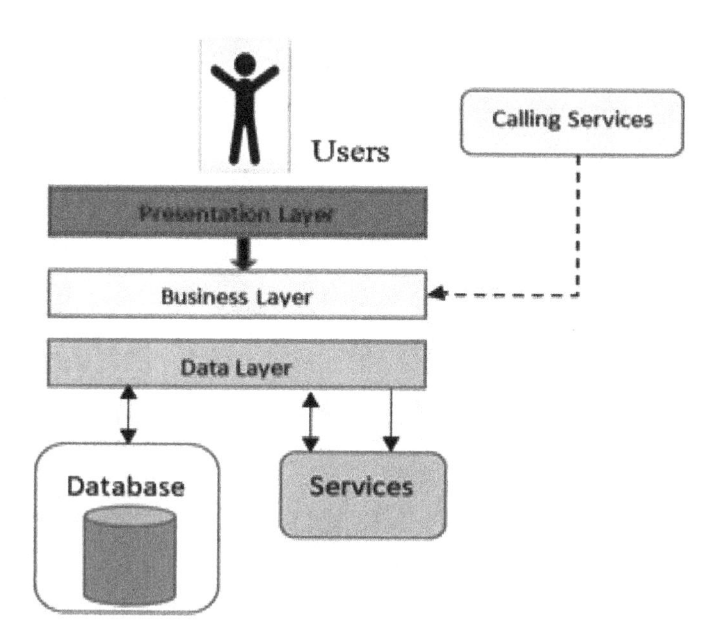

FIGURE 6.7 Mobile application architecture design.

designers successfully handle all the worst-case scenarios by properly placing the authentication, exception handling, data validation and verification, buffering, and app logs. The blocks of the business logic should be modular, and rules and policies need to be laid down for the app usage, access, and designing.

Data access layer:

The data layer works on the dataflow, accessing the data, components that facilitate data storage and management, etc. The data layer comprises utility tools of data, data-handling components, and service agents. The success of a project hinges on the development of a stronger application architecture [39, 40]. The data access layer ensures the safe data exchange by using the right data access technologies and secure and functional solutions. The layer can use shared preference, external storage, SQLite – table-based, cloud-based, and fire-based storage. While designing the app, the designer should think about the right kind of storage based on the type of the data to be stored and processed. Also, the designer should ensure the security of the data and seamless transition between the different layers.

6.6.2 Choosing the Architecture for Your Mobile Application

The designers need to select the right design architecture based on the client's requirements as shown in Figure 6.8. If there is a financial constraint, developing native programs is preferred, because they provide straightforward functionality and performance.

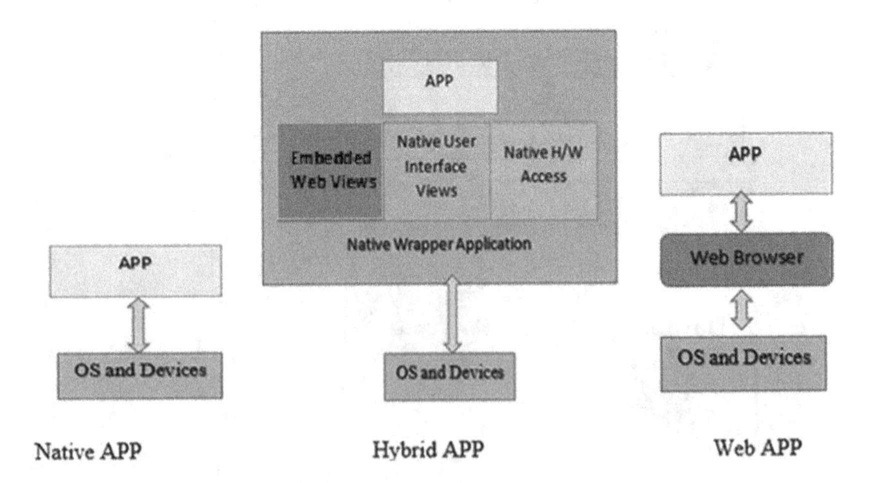

FIGURE 6.8 Mobile APP technology stack.

Hybrid Apps are preferred if you choose to build an app that uses an approach of "build once, run anywhere". A multi-platform framework needs to create cross-platform solutions that can provide good experience supporting wide range of mobile devices. The multi-platform architecture may be cost-effective if you target all operating systems including windows but need to compromise on availability.

Web-based applications can provide an increased visibility to the customers for supporting wide range of devices. Designer can think of combining both the native and online development, allowing better user engagement. While native applications can provide the finest user experience, web applications increase your company's visibility by giving customers the option of using a range of devices.

6.6.3 Apps' Success

Mobile app design success not only depends on the layers, architectures used, but also on other aspects as well. To design a successful mobile application architecture, the entire scenario of the business needs to be analyzed, and this is how you can establish future targets and design apps that can be scalable, sustainable, and meet the everyday needs of the customer.

Before you begin creating your app architecture, keep the following points in mind:

Determine the device type:

Figure 6.9 shows the various types of devices. There are many various sorts of smartphones; therefore it's crucial to assess the device and its features before deciding on an app architecture. The following device features should be kept in mind like resolution of the screen, size of the screen, features of CPU, space for storage for the required memory, and the ease of use of framework design.

While designing the apps for mobiles, all these device features need to be considered as the app's intended functionalities may necessitate the use of customized software and hardware.

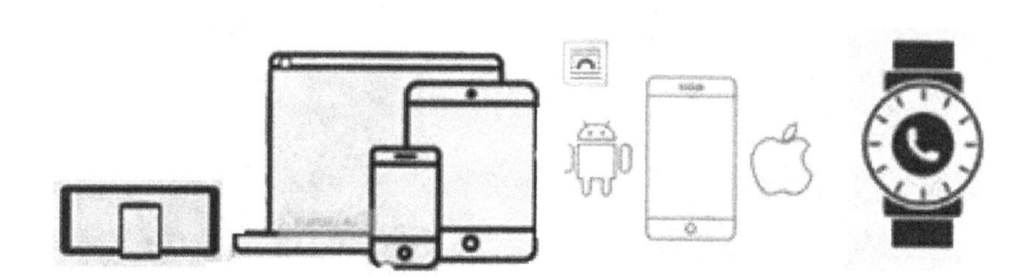

FIGURE 6.9 Different types of devices.

Bandwidth in mobile network:

It is critical to understand the online environments of the users to build the app, which are a complete picture of the Internet speed, type of the network used, and the location of the users. This will assist in building a robust app that can work well even in worst-case scenarios.

Also, the power consumption, battery lifetime, and the speed required by the app need to be analyzed before choosing software and hardware. During the failure you need to ensure that the state management, the data persistence, performance of the app is retained even with the bandwidth issues. Hence, you should think of these issues before hand and cache all the necessities.

UI design:

Creativity plays a vital role in expressing yourself artistically in your own way in creating an elegant UI design. While designing the UI, the designer needs to think about the ease of users' interaction with the app—the simpler the interface the better the user friendly feature. Nested interfaces or complex UI designs may lead to inappropriate functioning of the app.

Mobile application navigation:

Navigation in Mobile App is critical for navigating through the activities and data handling. Figure 6.10 shows an example for navigation. When

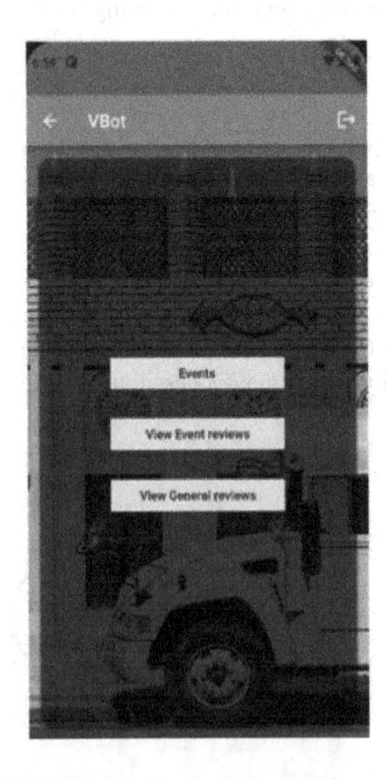

FIGURE 6.10 Navigation example.

deciding on a navigation approach, keep in mind both your personal tastes and the app's requirements. This is critical because it will significantly affect the user experience. You should evaluate and select the most appropriate navigation methods for your situation from the different options accessible. Some of the most popular ones include Navigation Drawer, Menus, Sticker Pack, Single View, and Scroll View.

Designer must adhere to precise standards when developing your app to ensure that it runs well in a variety of conditions.

6.7 APP DESIGN ENVIRONMENT

Numerous design problems exist which makes an application extremely complex and not easily accessible by a user. These issues can easily deem an app not practical. Designing for applications has been proven to be complex and challenging with no proposed guidelines for designing an effective and usable application that is scientifically proven. One of the most difficult aspects of building urban applications is determining "what activities contribute to communities" and "what tools can successfully assist people' activities".

However, there are eight golden rules of design, which are used as the standard for designing any application [41–42]:

- Unbiased
- Flexibility
- Simple and in-built
- Distinguishable information
- Fault tolerance
- Less technical and manpower effort
- Support
- Instructional climate

While designing any application it should be taken into account that the design should be simple and easy enough for students to understand without written instructions or skills to interact with [43].

There are two phases when it comes to the application design environment—analyzing and designing.

Analyse phase:
The designer determines the needs and the difference between the learners' current knowledge, abilities, and behaviors that they must have or are anticipated to have in this step. In basic words, a rudimentary need analysis is conducted, after which the roots of the problem are laid before the designer to solve.

Design phase:
The development strategy is defined in this step based on the data gathered during the analysis phase, and the method for achieving the goals is explained. To put it

another way, this is the section where the teaching technique, learning theories, learning objectives and activities, lesson plans, instructional strategy, and media selection are all made explicit.

When it comes to the opinion of the public on application design in terms of presentation and accessibility, it is noticed that design is an integral part of any working application. It indicates that the importance of user interface leads to a better working application. User interface design of any application is a key concern for any application's usefulness.

The term "environment" refers to the closed region which a situation and application cover a user. Ignoring the environment while designing the application will affect the overall experience. Analyzing the environment is of prime importance while designing any application.

Saving time of a user is a major factor while it comes to an application environment.

Let us take an example: "Swiggy" is a service-based application. It provides delivery-based services to users. People have the necessity to order food, and 'Swiggy' provides this service to its users. Food is delivered quickly, keeping the users satisfied who would always want to come back for more. This is an example of 'Swiggy' creating the perfect environment for its users.

Keeping a user longer in an environment which makes them use the application is a good thing. They have a requirement that can be solved using the XYZ application.

Environment plays a major factor while designing an application. Finding the perfect environment helps in providing a great and long-lasting impression for the user so that they always come back to use it.

The different environments available for Mobile App Development are:

A. Android Studio

Android Studio is a formal integrated development environment that assists in android app application development powered by IntelliJ. The Android Studio is bundled up with good code editor and strong developer tools that can provide the user with an ease of development of apps. Android is available with good configurations such as the following.

a. An MVC-based development environment for providing an effective app designing. An Android Studio snapshot [44] is shown in Figure 6.11.
b. The build system configuration done by Gradle.
c. An inbuilt emulator to preview the apps.
d. The running codes getting updated without restarting with apply changes.
e. Comprehensive framework with GitHub integration to load and use the templates.
f. Automated version controls and compatibility with all android OS versions.

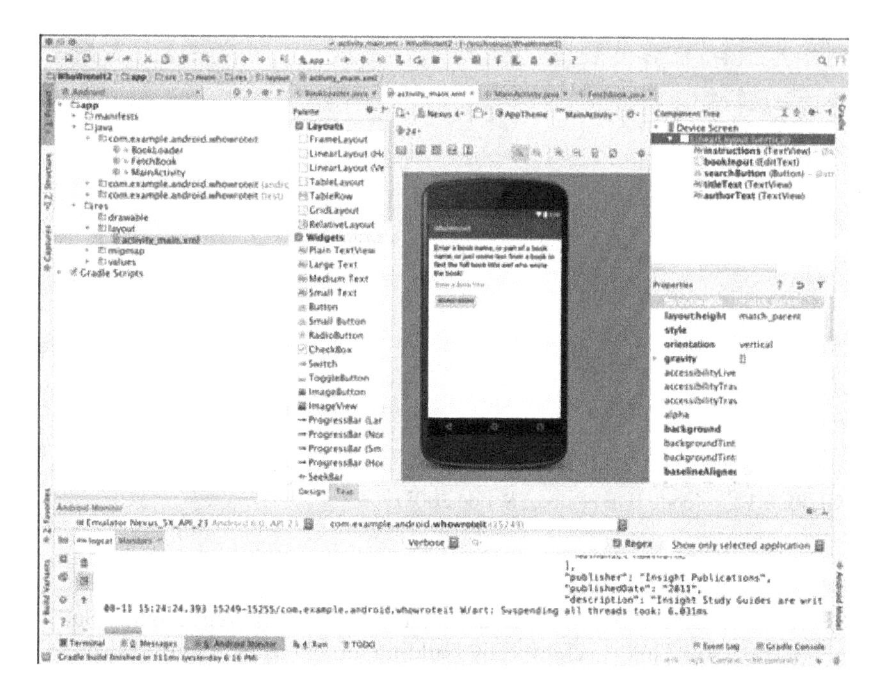

FIGURE 6.11 Android Studio for mobile application development (source: Android Studio 2019).

g. Programming support with C++ and the NDK.
h. Google firebase cloud integration to store and retrieve data related to the apps.

B. iOS

Apple Inc. created the iOS mobile operating system for iPhones, iPads, and other Apple mobile devices. After Android, iOS is the second most popular and widely used mobile operating system with the following features.

a. The iOS operating system is built on a layered framework.
b. The iOS does not allow communication between each layer directly.
c. The cocoa layer provides all the controllers, widgets, and all the main system functionalities.
d. The core services layer provides all the fundamental resources of the app. The lower-level layers provide essential services that all applications depend on.
e. The media layers provide graphics and user interface-related functions.
f. Xcode IDE [45] to design the apps is as shown in Figure 6.12.

FIGURE 6.12 Xcode IDE (source: Xcode 2003).

C. Cross-platform development

It refers to developing a single app that works on different mobile platforms [46]. This is achieved by building an app using a universal programming language such as JavaScript, for React Native, and Dart, for Flutter. Cross-platform development helps optimize the cost of development, time, and resources while increasing mobile apps' efficiency and performance. The best cross-platform app development tools [23, 47] are as follows:

a. Phone Gap (Apache Cordova)

Apache Cordova facilitates the development of apps on free cross-platform [48]. Figure 6.13 represents PhoneGap. PhoneGap is available for free, and it is an open-source platform developing cross-platform mobile apps, powered by Apache Cordova [49].

It has the inbuilt compiler and a debugger supported with testing tools. HTML5, CSS3, and JavaScript are among the most important web technologies included. PhoneGap is supported by a number of plugins, including Image Capture, Media Recorder, Push, and others. PhoneGap's plugins allow you to activate various smartphone functionalities such as location, accelerometer, camera, and sound.

In-app purchases are also supported by PhoneGap in both the iOS app store and the Google Play Store. It is a best tool for designers that know their way around front-end technologies. A third-party cache and graphics accelerator can help the application run faster.

Unique features of this tool are the following.

- A solid backend for novices
- Open source, so it's easy to share
- Extremely adaptable for developers

FIGURE 6.13 PhoneGap (Source: Apache Cordova 2009).

FIGURE 6.14 Appcelerator (source: Engadget 2011).

b. Appcelerator

Appcelerator [50] is one more tool for designing apps that support cross-platform. Figure 6.14 represents Appcelerator. This also checks the functionality of mobile app by creating, deploying, and verifying the mobile development tool [51]. This supports JavaScript as a universal coding mechanism.

It provides platform robustness by using the same code for several platforms, such as windows, iOS, Android, and Blackberry. An expert java developer can use this platform to quickly design the mobile application.

Appcelerator runs with no schema for the database, and this helps in deploying the data models with zero configuration. Salesforce, MS Azure, MongoDB, MS SQL, and Box are among the pre-built integrations. The latest version addresses the previous concerns with speed and lag.

Remarkable features are given as follows.

- It has a titanium framework, which makes it extremely cost-effective.
- Cloud-based services for packaging, distribution, and analysis
- The user can access the fundamental source code at any moment.

c. React Native

React Native is a Facebook's open-source tool for designing apps for cross-platforms. Figure 6.15 represents React Native. It adheres to the "learn once, write everywhere" principle. The basic language is JavaScript, which allows for native capabilities [52].

You may use the same base code to port your app to Android and iOS once it's been coded.

The capability that distinguishes React Native is that, although being cross-platform, apps developed on this platform have the appearance and feel of native apps, providing an engaging user experience.

It's a new framework that allows for rapid prototyping and high initial velocity. Furthermore, if you are skilled in JS, learning the basics and getting started are simple. While this sounds fantastic, the JavaScript code consumes a lot of RAM when performing calculations. It's an excellent choice for agile development, and, in many circumstances, proves to be more successful.

Hallmarks of ReactNative are given here.

- Makes your app more visible in app stores
- Uses powerful web technologies to provide a modular and intuitive architecture
- Is community-driven, resulting in excellent support.

d. NativeScript [53]

NativeScript is equivalent to writing native apps with cross-platform capability in JavaScript, TypeScript, or Angular.

It is the most preferred app development framework to create trending microapps. Microapps help you perform functionalities with precision. They usually focus on a single feature.

NativeScript is based upon the following major components:

- Runtimes
- Core modules
- CLI
- Plugins

FIGURE 6.15 React Native (Source: Storm 2015).

The runtimes allow the developer to interface the iOS and Android APIs using JavaScript code. Figure 6.16 represents NativeScript. Core modules help to summon the abstractions needed for the native API. The CLI command helps the user to create, run, and build apps using NativeScript.

The plugins form the building blocks of this app development framework. Like the plugins that help make mobile apps, it contains some functionality and aids the developer in building apps faster.

Core features of this tool are as follows.

- Allows use of a complete stack of cross-platform APIs.
- Provides "zero days" support for new native frameworks.
- Reach, reusability, and rapid app development.

e. Flutter

Flutter [54] is known for its speedy and dynamic app development environment for the cross platforms. Figure 6.17 represents Flutter developed by Google and released in 2017. The Flutter files are executed

FIGURE 6.16 NativeScript (source: NativeScript-Vue 2018).

FIGURE 6.17 Flutter (source: Google 2017).

on a Dart Virtual Machine (DVM). In Flutter, a widget tree is generated automatically with the changes that were made in the design. The Flutter widget contains the views, icons, scroll views, navigations, fonts, etc. to support both iOS and Android app.

Flutter works like a React framework, with inbuilt 2D for widgets, tools, and rendering. This provides an ease for the developers to design, test, verify the app.

Hallmarks of Flutter are as follows:

- Productive as developers can build iOS and Android with same code.
- Material design helps in building customized app.
- Creating brand-driven designs with OEM widgets.

f. Xamarin

Xamarin [55] is a cross-platform programming framework for mobile, desktop, and tablet devices. Figure 6.18 represents Xamarin. The Android and iOS quick-starts assist developers in creating native UI components. C#, F#, and the Razor template engine are all supported by Xamarin.

In February 2016, Microsoft acquired Xamarin and later released the Xamarin SDK. A free tool to integrate enterprises' features using Microsoft Visual Studios is also available.

The native friendly interface empowers developers to respond to any specific client demands. It is a one-size-fits-all program with HTML tools to help developers and end users.

Primary attributes are as follows:

- Because the new version has a smaller disc footprint, it installs more rapidly.

FIGURE 6.18 Xamarin (source: Microsoft 2016).

- Automatic Android SDK installation is based on requirements.
- Offers fantastic visual experience for iOS and Android.

g. Felgo

Felgo [56] is an easy-to-use cross-platform development SDK which works on the Qt framework. Qt, a famous C++ structure, is widely used by developers. Felgo enhances this Qt core with an app and game element. Analytical tools, plugins for in-app purchases, etc., make it a standalone platform. Figure 6.19 represents Felgo, a cross-platform development tool.

Felgo is one of the growing cross-platform Mobile App Development tools. It is today very popular with the new generation of developers. The feature that makes it attractive is that they won't charge you if you are not earning profit with your business.

Felgo holds modern developer's attention because as compared to Xamarin and ReactNative, they can save 90% of coding with Felgo. It is moving towards the top spot in providing a secure, time-saving, and beginner-friendly platform with adequate support.

Features that make Felgo unique are as follows.

- Faster and stunning native, cross-platform apps with excellent UIs
- Creates powerful embedded systems app using C++ and JavaScript
- Develops games faster with an in-game level editor, multiplayer, and Felgo cloud services
- Performs live reloading of codes, at one, reloads per second from Windows, Linux, and Mac

FIGURE 6.19 Felgo (source: Felgo 2019).

h. Ionic

To create hybrid apps, the Ionic Mobile app Development framework [57] employs HTML5. When it comes to support, speed, and third-party access, hybrid apps outperform pure native apps. If you just have a short amount of time to market your software, this is the ideal alternative.

You can have native-style UI components and layouts on iOS and Android. However, because it is an HTML5 framework, it requires a wrapper such as Cordova or PhoneGap in order to function as a native app. Figure 6.20 shows Ionic, an open-source SDK.

Recent changes are listed as follows.

- Stencil is a web component compiler that aids in the creation of standards-compliant web designs.
- Capacitor—a layer, also known as cross-platform API, that allows web programmers to call native SDKs.
- Lazy Loading—comes with Ionic 4 and allows you to divide your app into bundles to speed up loading time and performance.

i. RhoMobile

RhoMobile [58] provides Rhodes, an open-source Ruby framework that works with Android, Windows Mobile, Symbian, iPhone, and RIM. Ruby is a well-liked programming language. Figure 6.21 shows RhoMobile, which an open-source framework. RhoMobile comes with RhoHub, a hosted development environment for developers to write the app, and RhoSync, an independent server that helps you store all of the app's data and keep it up to date on the users' devices.

You may create native-like programmes that run on multiple platforms using a single code base. Native programs do a fantastic job of working with the hardware that is available. As a result, the job is completed more efficiently, quickly, and accurately.

FIGURE 6.20 Ionic (source: Ionic 2013).

FIGURE 6.21 RhoMobile (source: RhoMobile 2017).

FIGURE 6.22 Kony Mobile cross platform (source: Kony 2009).

Advantages of RhoMobile are as follows:

- Uses Ruby language which is simple and productive
- JRuby allows Ruby Apps to run on Java containers
- The DRY approach avoids the repetition of codes enabling smooth development

j. Kony Mobile Platform

Kony Mobile Platform provides all the essential functionalities to make universal apps. The channel-specific APIs, third-party libraries using FFI (Foreign Function Interface), etc. are its specialties. Figure 6.22 represents Kony Mobile cross platform.

The Kony One Server application server further enables offline sync, security, detection of a device for mobile-optimized content, SMS, push notifications, etc.

The Kony Visualizer powers the front end with the multi-channel JS API framework and assists designers and developers to deliver an array of services to smartphones, tablets, and desktops on iOS, Android, Windows, and Blackberry.

Essential qualities of Kony Mobile Platform are as follows:

- Import capability for any Native API from Visualizer
- Allowing the use of the open-source framework for Jquery, Bootstrap, and Angular
- Local packaging of web content in the app and native to web contexts

k. Sencha Touch

In terms of functionality, Sencha's [59] ExtJS is similar to React Native. The use of the Ext JS framework for development is a key difference between the two. Figure 6.23 represents Sencha Touch. It has a visual HTML5 application builder as well as the ability to reuse custom components. The native application packager is in charge of Apple Store distribution.

You can test and execute your applications on both browsers and mobile apps because it uses HTML5 for development. It can also be used to create fully functional apps using cross-platform tools such as Adobe PhoneGap.

Characteristics of Sencha Touch are as follows:

- Enhances development team's productivity through IDE plugins
- Minimizes web app development cost
- MVC architecture to make codes readable

Cross-platform app making is a turning point in the Mobile App Development field. The Internet revolution demands your business to have an online presence. More and more entrepreneurs are shifting their business online.

You can engage customers with stunning cross-platform apps like PIU, Peekay Steels, and C3 Card.

Cross platform

FIGURE 6.23 Sencha Touch (source: Sencha Touch 2017).

6.8 APP DEVELOPMENT PROCESS

Let's start talking about Mobile App Development. Over the last decade, a lockdown was imposed all around the globe. Since that time to till now, people started working from home and got used to use mobile apps for ordering food and groceries. Each mobility solution is a huge part of our daily lives. In India alone, consumers are spending 4.6 hours per day on mobile devices. In recent times, mobile app has drastically increased its requirement, generating billions of dollars' revenue via app stores such as Play Store. In most recent times, companies are trying to take advantage of this trend. But the major drawback is that most of the people don't know how to create a successful mobile app with a proper function and run in the market. In this document, let's see how the mobile apps can be developed and worked on in a proper manner.

Steps involved [50] in app development are as follows:

- Strategy
- Analyse and plan
- UI/UX design
- App development
- Testing
- Deployment and Support

6.8.1 Strategy

In this development process, in order to successfully run your app, there needs to be a specific way or an idea to implement the project. You need to decide what needs to be done for your app to run successfully, as the strategy differs from one app to another as per their requirements, but the minimum requirements for every app are the following:

- Know who are the users
- Do adequate research on the app
- Know the goals of the project
- Select a specific platform

6.8.2 Analysis and Planning

In every app development, this is the most important part because here is where the app starts taking shape with appropriate use-cases and getting full details of the functional requirements.

By this, you prioritize the product requirement and group the requirements into specified steps. Coming onto this part, you need to know the skills that are needed for an app to develop and to make sure for whom are you developing it like as Android users or IOS users. For that you need to include iOS or Android Developers. The most important part is selecting the name for the app which should be unique, and you should be able to guess or have a hint of its function just by reading that name, and there needs to be some research done for this.

6.8.3 UI/UX Design

This step plays a major role to deliver smooth and comfortable experience to a user. Every successful app depends on the UI /UX design which is provided to each and every user around the globe. Creating great user experiences such as being interactive, user-friendly, with great finishing of the app makes it look more beautiful and attracts users towards the app.

Start making UI/UX designs by planning it on paper and create a specific layout on each page that you need to develop into a proper structure that meets all functional requirements. While designing, do it in a proper way because each and every user might not have the same device that you are using so design carefully where all the devices meet the requirements and provide a great experience to the user.

6.8.4 App Development

This part remains the integral part before the real effort starts. First of all you need to define the technical part, identify all the technology need to be used, and start ticking up milestones.

In this, database integration (*recommended Google firebase) is included, and you need to use proper API (Application Programming Interface) for connecting the app and the server/database.

You also need to keep in mind the requirements of end users and whether they need Internet to access the app or not and accordingly you need to design an app. iOS Apps can be developed using Objective-C or Swift programming language, and Android Apps can be developed using Java/Kotlin.

As each development milestone gets completed, it is passed on to the app testing team for validation.

6.8.5 Testing

As performance matters everywhere, people want the app to run smoothly, and it should be user friendly. There need to be some people to keep finding the bugs to perform a trial and error method to check the software quality and keep on track for checking the errors.

Some steps for the testing process are:

- Functionality
- Performance
- Security
- Platform

6.8.6 Deployment and Support

In order to release the app to the app store or play store, you need to have a developer account to launch it. Once you submit it to the Play Store, the app undergoes review, as the app will be available within hours, but it isn't the case with iOS. You need to go over it thoroughly and encourage users to submit feedback and suggestions for your app to your company. End-user support and frequent app updating with enhancements will be critical for keeping users engaged. Updates to mobile apps must go through the same submission and review process as the original. Furthermore, with native mobile apps, you must keep up with technological changes and regularly update your app for new mobile devices.

6.9 CONCLUSION

This chapter has provided the comprehensive details of design thinking aspects related to the mobile application development. This chapter has provided the guidelines to be followed while structuring the layers, the architectures, the design environments, customer satisfaction, prototyping, and app designing.

ACKNOWLEDGEMENTS

I would like to appreciate the efforts of my students Abhay Chaudhary, Tanaya Krishna Jupalli, Sneha V., N. Shashank Varma, Teki Gupil Venkata Sai, and Ch Sai Sumedh for sharing their inputs to bring out this chapter.

REFERENCES

1. Dell'Era, C., Magistretti, S., Cautela, C., Verganti, R., & Zurlo, F. (2020). Four kinds of design thinking: From ideating to making, engaging, and criticizing. *Creativity and Innovation Management*, 29(2), 324–344.
2. Gero, J. S., & Milovanovic, J. (2020). A framework for studying design thinking through measuring designers' minds, bodies and brains. *Design Science*, 6.
3. Meinel, C., & Leifer, L. (2012). Design thinking research. In *Design thinking research* (pp. 1–11). Springer, Berlin, Heidelberg.
4. Brown, T. (2008). Design thinking. *Harvard Business Review*, 86(6), 84.
5. Gobble, M. M. (2014). Design thinking. *Research-Technology Management*, 57(3), 59–62.
6. Liedtka, J. (2018). Why design thinking works. *Harvard Business Review*, 96(5), 72–79.
7. Suzianti, A., Wulandari, A. D., Yusuf, A. H., Belahakki, A., & Monika, F. (2020). Design thinking approach for mobile application design of disaster mitigation management. In *Pervasive health: Pervasive computing technologies for healthcare* (pp. 29–33). ICST. https://doi.org/10.1145/3379310.3379324.
8. Hsu, T. H., Horng, G. J., & See, A. R. (2021). Change in learning motivation observed through the introduction of design thinking in a mobile application programming course. *Sustainability (Switzerland)*, 13(13). https://doi.org/10.3390/su13137492.

9. Aulia, N., Andryana, S., & Gunaryati, A. (2021). Perancangan user experience aplikasi mobile charity menggunakan metode design thinking user experience design of mobile charity application using design thinking method. *Jurnal SISFO-TENIKA*, 11(1), 26–36.
10. Schiele, K., & Chen, S. (2018). Design thinking and digital marketing skills in marketing education: A module on building mobile applications. *Marketing Education Review*, 28(3), 150–154. https://doi.org/10.1080/10528008.2018.1448283.
11. Alimin, S. R., Hamid, N. H. A., & Nasruddin, Z. A. (2020). City i-Tick: The android based mobile application for students' attendance at a university. *Bulletin of Electrical Engineering and Informatics*, 9(3), 1284–1290. https://doi.org/10.11591/eei.v9i3.2078.
12. Hou, I. C., Lan, M. F., Shen, S. H., Tsai, P. Y., Chang, K. J., Tai, H. C., . . . Dykes, P. C. (2020). The development of a mobile health app for breast cancer self-management support in Taiwan: Design thinking approach. *JMIR MHealth and UHealth*, 8(4). https://doi.org/10.2196/15780
13. Kamran, R., & Dal Cin, A. (2020). Designing a mission statement mobile app for palliative care: An innovation project utilizing design-thinking methodology. *BMC Palliative Care*, 19(1). https://doi.org/10.1186/s12904-020-00659-1
14. Sunder M, V., Mahalingam, S., & Krishna M, S. N. (2020). Improving patients' satisfaction in a mobile hospital using Lean Six Sigma—a design-thinking intervention. *Production Planning and Control*, 31(6), 512–526. https://doi.org/10.1080/09537287.2019.1654628
15. Islam, M. N., Islam, I., Munim, K. M., & Islam, A. N. (2020). A review on the mobile applications developed for COVID-19: An exploratory analysis. IEEE Access, 8, pp. 145601–145610.
16. Makalalag, A. H., Ekawardhani, Y. A., Valentina, T., & Gaol, L. (2021). User interface/user experience design for mobile-based project management application using design thinking approach. *International Journal of Education, Information Technology and Others*, 4(2), 269–274.
17. Jabangwe, R., Edison, H., & Duc, A. N. (2018). Software engineering process models for mobile app development: A systematic literature review. *Journal of Systems and Software*, 145, 98–111.
18. Adrian, F. (1999). The saving and spending habits of young people. *Journal of Economic Psychology*, 20(6), 677–697. ISSN 0167–4870.
19. Kumar, P. (2017, February). Impact of earning per share and price earnings ratio on market price of share: A study on auto sector in India. *International Journal of Research Granthaalayah*, 5(2), 113–118.
20. Guo, L., Sharma, R., Yin, L., Lu, R., & Rong, K. (2017). Automated competitor analysis using big data analytics: Evidence from the fitness mobile app. *Management Journal*, 23(3), 735–762. https://doi.org/10.1108/BPMJ-05-2015-0065
21. Dou, W., Lim, K. H., Su, C., Zhou, N., & Cui, N. (2010). Brand positioning strategy using search engine marketing. *MIS Quarterly*, 261–279.
22. Roach, C. (2006). Thought leader: Harness your unique selling point. *Strategic Communication Management*, 10(3), 5.
23. Rahman, M. R., & Safeena, P. K. (2016). Customer needs and customer satisfaction. Retrieved from GitHub: https://github.com/
24. Hoehle, H., & Venkatesh, V. (2015). Mobile application usability: conceptualization and instrument development. *MIS Quarterly*, 39, 435–472. doi: 10.25300.MISQ/2015/39.2.08. [CrossRef][Google Scholar].
25. Attili, V. P., Mathew, S. K., & Sugumaran, V. (2021). Information privacy assimilation in IT organizations. *Information Systems Frontiers*, 1–17.

26. Chatterjee, S., Chaudhuri, R., Vrontis, D. and Hussain, Z. (2021). Usage of smartphone for financial transactions: From the consumer privacy perspective. *Journal of Consumer Marketing*. doi: 10.1108/JCM-03-2021-4526 (ISSN: 0736-3761 – Emerald).
27. Sharma, T., Dyer, H. A., & Bashir, M. (2021). Enabling user-centered privacy controls for mobile applications: Covid-19 perspective. *ACM Transactions on Internet Technology (TOIT)*, 21(1).
28. Sheung, C. T., Yazdanifard, R., & Park, F. B. (2014). How online privacy systems assure more customer satisfaction and trust?. *International Journal of Economic and Management Science*, 3(2), 18–21.
29. Sadler-Smith, E. (2015). Wallas' four-stage model of the creative process: More than meets the eye? *Creativity Research Journal*, 27(4), 342–352.
30. Amabile, T. M. (1983). The social psychology of creativity: A componential conceptualization. *Journal of Personality and Social Psychology*, 45(2), 357.
31. Boden, M. A. (2009). Computer models of creativity. *AI Magazine*, 30(3), 23–23.
32. https://appinventiv.com/blog/top-mobile-app-prototyping-tools/
33. www.usability.gov/how-to-and-tools/methods/prototyping.html
34. Figma: the collaborative interface design tool. www.figma.com
35. InVision: www.invisionapp.com
36. www.adobe.com/products/xd/solutions/app-design-software.html
37. https://azure.microsoft.com/en-us/get-started/azure-portal/mobile-app/
38. Sketch: www.sketch.com
39. Park, C. H. and Kim, Y. G., 2003. Identifying key factors affecting consumer purchase behavior in an online shopping context. *International Journal of Retail & Distribution Management*, 31(1), 16–29.
40. Strauch, S., Andrikopoulos, V., Breitenbücher, U., Sáez, S. G., Kopp, O., & Leymann, F. (2013). Using patterns to move the application data layer to the cloud. *Proceedings of PATTERNS*, 13, 26–33.
41. Muslimin, M. S., Nordin, N. M., Mansor, A. Z., & Yunus, M. M. (2017). The design and development of MobiEko: A mobile educational app for microeconomics module. *Malaysian Journal of Learning and Instruction*, 221–255.
42. Sasao, T., & Konomi, S. I. (2014, June). U. App: An urban application design environment based on citizen workshops. In *International Conference on Distributed, Ambient, and Pervasive Interactions* (pp. 605–616). Springer, Cham.
43. Jayaram, S., Jayaram, U., Wang, Y., Tirumali, H., Lyons, K., & Hart, P. (1999, November–December). VADE: A virtual assembly design environment. *IEEE Computer Graphics and Applications*, 19(6), 44–50.
44. https://developer.android.com/studio/run
45. https://developer.apple.com/xcode/ide/
46. Grace Chu, Okan Arikan, Gabriel Bender, Weijun Wang, Achille Brighton, Pieter-Jan Kindermans, Hanxiao Liu, Berkin Akin, Suyog Gupta, and Andrew Howard. Discovering multi-hardware mobile models via architecture search. *arXiv preprint* arXiv:2008.08178, 2020.
47. Palmieri, M., Singh, I., & Cicchetti, A. (2012, October). Comparison of cross-platform mobile development tools. In *2012 16th International Conference on Intelligence in Next Generation Networks* (pp. 179–186), IEEE, IDT, Malardalen University.
48. www.javatpoint.com/cordova-phonegap
49. Ghatol, R., & Patel, Y. (2012). *Beginning PhoneGap*. Apress Media, New York.
50. Işitan, M., & Koklu, M. (2020). Comparison and evaluation of cross platform mobile application development tools. *International Journal of Applied Mathematics Electronics and Computers*, 8(4), 273–281.

51. www.engadget.com/2011-09-19-appcelerator-launches-app-store-for-app-components.html
52. https://semaphoreci.com/blog/what-is-react-native
53. Anderson, N. J. (2016). *Getting Started with NativeScript.* Packt Publishing Ltd. Birmingham.
54. Wenhao Wu (2018). *React Native vs Flutter, Cross-Platforms Mobile Application Frameworks,* Thesis at Metropolia University of Applied Sciences.
55. Rasmusson Wright, Y., & Hedlund, S. (2021). *Cross-platform Frameworks Comparison: Android Applications in a Cross-platform Environment. Xamarin Vs Flutter. Dissertation,* Blekinge Institute of Technology, Sweden.
56. Aljzaere, H., Khan, O., & Hardt, W. (2021) *Adaptive User Interface for Automotive Demonstrator.,* Master thesis at Chemnitzer Informatik-Berichte.
57. Sharma, S., & Kumar, P. (2020). Rest house hybrid mobile application using ionic tool. *International Research Journal of Engineering and Technology,* 7(4), 45–50.
58. Hudli, A., Hudli, S., & Hudli, R. (2015, October). An evaluation framework for selection of mobile app development platform. In *Proceedings of the 3rd International Workshop on Mobile Development Lifecycle* (pp. 13–16). Association for Computing Machinery, New York, NY.
59. Garcia, J., De Moss, A., & Simoens, M. (2013). *Sencha Touch in Action.* Manning Publication Co., Croatia.

7 Design Thinking – Networking and Telecommunications, a Review

A. Kishore Kumar[1], *P. K. Poonguzhali*[2] *and T. Nivethitha*[2]
[1]Department of Robotics & Automation, Sri Ramakrishna Engineering College, Coimbatore, India
[2]Department of Electronics and Communications Engineering, Hindustan College of Engineering and Technology, Coimbatore, India

CONTENTS

7.1 INTRODUCTION

Design thinking has piqued the interest of practitioners and academics alike, as it provides a fresh approach to problem solving and innovation in the networking and telecommunications fields. Design thinking is an analytical methodology which can acquire a solution for problem definition [1] by understanding the

DOI: 10.1201/9781003189923-9

customer needs involved, re-framing the quandary in human-centric ways by creating many new ideas, and taking a concrete approach in prototyping and trying effectively for handling complicated problems that are imprecise or unidentified. We can be successful in overcoming the difficult challenges that we face in day-to-day life using the design thinking methodologies by discerning the five steps of design thinking such as understand, classify, ideate, model, and test. As per the statement of Tim Brown, Chair of IDEO, "*design thinking is a human-centered approach to innovation that employs the designer's toolset to blend people's wants, technology possibilities, and economic success requirements*". He constantly emphasizes the importance of design thinking. Opinion from designers may transform the business fabricate products related to the services processes and tactics to build. This systematic approach helps in design thinking that merges the desire from a customer viewpoint and the availability of practical technology with feasible cost-effectiveness. It also facilitates individuals who aren't designers to use inventive tools to work out a variety of problems.

The first stage of design thinking process is to provide solution to any problem by procuring an empathic knowledge to address. This involves discussion with trained people to find out more about the area under discussion by examining, appealing, and understanding with people to better comprehend their knowledge and inspiration, as well as submerge toward a substantial situation with individual perceptive of the issue at hand. Empathy is necessary in a human-centered design approach ever since it let designers to place their individual world vision aside for getting a standardize insight into needs of clients [2]. Owing to the time limitations, a considerable amount of data is composed at the first stage which can be used in the subsequent stages so as to set up the best possible perceptive of the clients, their stress, and the complexity that can strengthen the development of that precise product. The information that was extended and collected during the Empathize stage is placed together during the Define step. In this stage, we must examine and synthesize the observations in order to define the key difficulties that have been observed. The problem should be expressed as a predicament statement in a client-centered approach at this point.

During the third stage, designers are geared up by bringing up their ideas during design thinking process. After empathizing with user's needs in the Empathize stage, analyzing and synthesizing what was observed in the define stage, designers have to come up with a client-centered problem statement. With this solid foundation, new optimized solutions to the problem statement must be established by finding different ways of looking at the setback (thinking outside the box). At the beginning of the ideation phase, it's vital to collect as many ideas or issues with different solutions as feasible. Different ideation strategies might be applied by the end of the ideation phase to determine the best way to solve the problem or provide the basics to come.

Different ideation strategies can be employed toward the end of the Ideation phase to determine the best solution to solve the problem or provide the necessities to come out of the dilemma. The model must be developed in the subsequent stage of the design thinking process, which is an experimental phase. Design team

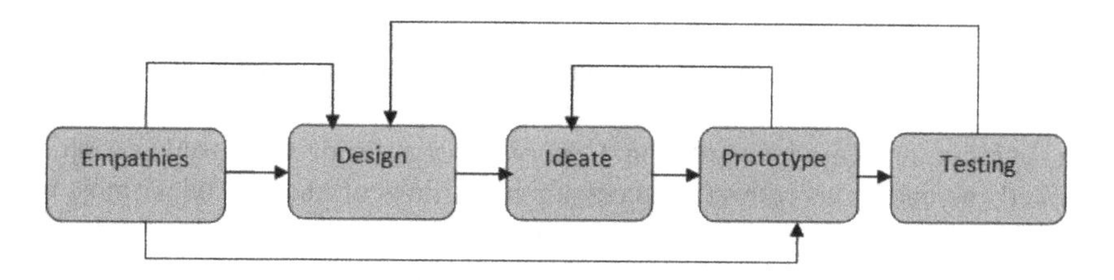

FIGURE 7.1 Design thinking: a five-step procedure.

must create and include different cost-effective, scaled-down versions of the product or specific aspects discovered inside the product so that the problem solutions created in the earlier phase can be investigated. Trial products can be tested within the product design team or in research and development departments or can be tested by testing department engineers who are not part of the planning team [3]. The objective of the investigational phase is to discover the finest resolution for each of the challenges proposed in the previous three stages. The responses are implemented in the model in a step-by-step process based on the users' knowledge, and they are approved, improved, and re-examined, or rejected. By the end of this phase, the design team will have a better understanding of the product's limitations and issues, and also through indulgent of how real users will perform, reflect, and experience when relate with the finishing product. The designers or evaluators extensively test the whole product by making use of the best ideas developed during the prototyping development in the last step of design thinking. The results collected during the testing phase are often used to redefine one or more problems and to educate the understanding of the users; the conditions of usage; how people think, behave, and feel; and to empathize in an iterative process. Even at some stage, changes and improvements are done to rule out potential problem solutions and gain a thorough grasp of the product and its users. Figure 7.1 illustrates the methodical processes of design thinking.

The following are some of the methods used by design thinking to solve complicated challenges.

- Empathizing: Distinguishing the human needs at hand.
- Characterizing: Reframing and defining the situation from a human perspective.
- Ideation: During ideation gathering, many ideas are generated.
- Prototyping: Using a hands-on approach to prototype development.
- Testing: Creating a problem-solving prototype that can be tested.

7.2 USING DESIGN THINKING TO IMPROVE PRODUCT SUPPORT FOR TELECOM SOFTWARE

In the context of telecom software, product support implies that customer complaints received via cell phone sustaining lines, classification systems, or additional means are primarily examined through the Enterprise Support Team in a typical Telecom Tech Product Support approach. The complaints are then rectified either by direct client consultations or by routing them rear to the back-end engineering team for resolution. Legacy Telecom Products, as well as sophisticated and enormous codes (in succession in millions of lines of codes), must be patched and sanity-feature-tested for many multifaceted scenarios ahead of being deployed out to client networks for any issue resolution.

7.2.1 Design Thinking in the Context of Telecom Software Product Support

According to Harvard Business Review, design thinking's basic concepts are empathy for users, prototyping discipline, and, last but not least, tolerance for failure. Owing to the hugeness of nature, the demand of an enhanced logging system, the number of coexisting users, and the number of parallel failures, the three concepts of empathy, prototype, and tolerance take part in a vital part in product support for telecom software. As the desire for improved experiences grows, numerous telecom businesses are incorporating design thinking into their software development processes.

The types of issues that were reported and how long it took to determine them from the end-user perspective must be used as feedback for future feature development, and hence similar issues can be resolved more quickly. When it comes to prototypes, they promptly reassure the customer that if the software system is given the appropriate inputs, something feasible may be accomplished. Giving the end user a range of options for addressing issues based on analytics that look at the types of problems that have happened and the solutions that have been offered helps improve tolerance for failure. The backend team can work more effectively if the basic solutions have been tried from the end-user perception [4].

7.2.2 Key Opportunities for Improvement at Diverse Stages

The telecom manufactured goods support scrap submission practice begins with the issue identification and isolation phase, which is followed by the support engineer's remedy proposal and implementation, and subsequently liberates level testing of multiple characteristic interface understanding levels, regression, and stress testing. A few of the areas where the telecom business should improve are given as follows:

On Tech Support stage:

a. Predicting services before they are required.
b. Assuring a reduction in invalid customer escalations.
c. Enabling smarter ticket routing, clustering, and assignment to the most appropriate source.

Next to Engineering Support stage:

a. Making sure faster triage and bug fix localization and suggestion
b. With automation and innovation, improving issue isolation and resolution
c. To avoid team disagreements, delegating the problem to the person with the appropriate competence.

By the Side of Testing/Certifying the Product Patch-Release stage:

a. Solving the activity problem and testing in a restricted set of conditions
b. Progressing the resolution accuracy by validating the modifications for this patch by executing the fixed test cases, accompanying scripts, environment, and test data.

Telecommunications can look for improvements in engineering support with the use of design thinking by incorporating the following aspects:

1. Working together as a group
2. A thorough understanding of the complete system, which can aid in determining the overall scenario.
3. When developing software, maintaining that the age of unit testing, integration testing, and performance testing in the lab will come to an end once the code is deployed in a real-world setting. The program is scaled up in terms of the number of users, concurrent calls, and area switchovers, among other things. As a result, during the software design cycle, it is necessary to consider how to make the logging system more effective. Engineers benefit from brainstorming as it helps them come up with new ideas.
4. Cognitive product support: Analytics can aid with billing, contracts, performance, and service assurance in this area. It can also aid in the immediate implementation of a solution once an issue has been reported. However, efficient analytics necessitates the use of adaptive software.

The idea is to exchange knowledge and spread it to many teams, hence increasing the knowledge base. It aids in making the test more varied, hence increasing the software's robustness. It educates individuals at the very beginning of their tech support careers, assisting them in becoming well-known with the product and allowing them to route more effectively and only when necessary. To end with, if customer equipment is rationalized with an issue-centric repository and probable outlook possibilities, the adaptive software design from previous experiences enhances customer happiness [5].

7.3 DESIGN THINKING FOR TELECOM ENGINEERING

Speedier and higher-quality fiber deployment would result in an improved network infrastructure, resulting in super-fast and stable connectivity that fulfils consumer performance demands and enables faster adoption of high-speed specifications.

7.3.1 Telecom Fiber Optic Cable Market in India

Fiber optic expertise can now be applied and used in a large amount of areas, including medical fields, military fields, telecommunication aspects, industrial sectors, data storage aspects, networking fields, and broadcast industries. And, because of the great demand, it is true to say that the telecommunications industry is the most prominent user of optical fiber technology. This claim has been increasing in India, driven by factors, for instance the growth of fixed broadband, the replacement of BSNL's "getting-old" copper network with fiber optic cable, the construction of an alternating network for defense, the upgrading of wireless backhaul networks, and the want for higher speeds by 3G and broadband wireless access networks. India has an entrenched backbone network that connects states and the center. Even as the FOC networks exist up to the block-level, the backhaul network for the Optical Fiber Cable Network has yet to be implemented. Nearly 80% to 90% of tower backhaul connections still use microwave links, which do not allow increased bandwidth capacity. The beauty of optical fiber is that its bandwidth capacity is practically limitless. Aside from increased rural penetration, spectrum scarcity is likely to worsen, resulting in an increased demand for FOC for backhaul and BTS access [6–7].

7.3.2 The FOC Market

Many residential and business developments throughout the world are requiring fiber-optic technology and wireless data connection solutions. It's part of the market's evolution and is the result of the alternatives becoming more sophisticated. Telecom cables are essential across the backbone and access networks. Applications such as Fiber to the Premises (FTTP), Fiber to the Home (FTTH), and Fiber to the Building (FTTB) are few of the primary factors boosting the demand for FOC globally. In India, FTTH/FTTB has been implemented and is a hastily expanding phenomenon for reliable communication networks.

7.4 A Network Scenario for Indian Telecom

India has a well-established backbone network that connects states and the center. Although the FOC networks exist up to the block level, the backhaul network for the Optical Fiber Cable Network has yet to be implemented. Nearly 80% to 90% of tower backhaul connections still use microwave links, which do not allow better bandwidth capacity. The beauty of fiber is that its bandwidth capacity is practically limitless. Aside from increased rural penetration, spectrum constraint is projected to worsen, resulting in an increased demand for FOC for backhaul and BTS access.

7.4.1 The Purpose of Design Thinking in Telecom

To solve a problem like connection and update ageing infrastructure, a change in the process and strategy to building new telecommunications networks is required. Design thinking, a technique of problem-solving that questions conventional assumptions and solves challenging problems through iteration and development, has helped a company like Biarri Networks to meet this demand for innovation.

This shift in mindset makes it possible to design and build more broadband and 5G networks in a relatively short amount of time. Network planners, designers, and organizations like Biarri Networks must use design thinking to consider more data up front before beginning projects. Changes to designs, equipment or network architecture shifts, new legislation, geographical surprises, or other unanticipated obstacles can all be adjusted faster than in previous decades [8].

The method of manual network design: Despite the high-tech possibilities that a lightning fast network can bring, Fiber network design has always been a laborious and manual procedure. Designers would make actual sketches on maps and satellite pictures, noting changes in text documents, spreadsheets, or on paper. After learning about design thinking, they realized there has to be a better way. Many of the parameters, or input data, stay consistent throughout all broadband networks, despite the addition of many additional variables and unexpected circumstances. Every network employs the same Fiber optic connection [9]. Whether it's trenching for an underground network or leveraging existing poles for aerial spans, we only have a few alternatives for how we put the cable. Figure 7.2 illustrates the evolution of Fiber optic Transmission and Networking in the 5G Epoch [10].

The advantages of the manual technique were obvious: designers had complete control over their data. They were able to locate every Fiber cabinet because

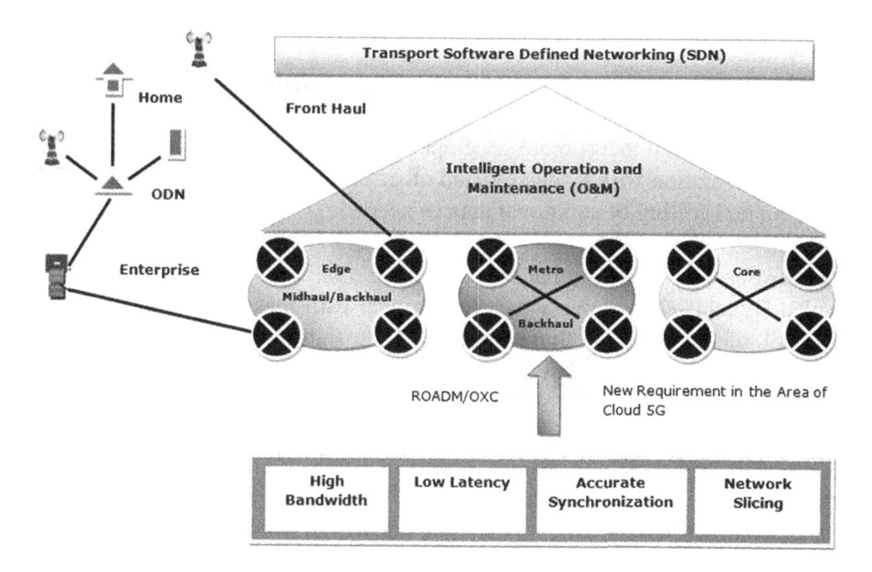

FIGURE 7.2 Fiber optic transmission and networking evolution in the 5G epoch.

they had plotted them on a map. Their extensive technical knowledge provided them with a comprehensive understanding of how networks may be centralized or spread. This highly technical skill did not transfer well to larger networks when these networks grew in size. On a single project, ten separate engineers could come up with ten distinct designs. Furthermore, no single network designer is capable of calculating the most cost-effective implementation manually. Consistency across networks is difficult to maintain with simply human controls. New input data could throw entire plans off, causing highly anticipated broadband building projects to be delayed unnecessarily. For Fiber networks, we're introducing machine-assisted auto design. How could we both enable designers and planners scale network building while yet preserving control of the particular demands of each project, using a design thinking approach to this network design problem?

We knew we wouldn't be able to match network designers' skill even if we had all the GIS data and mathematical inputs in the world. We weren't able to fully automate designs that had to respond to real-world issues. Artificial Intelligence and Machine Learning, unlike what we see in movies and hear on the news about AI taking over occupations, are not substitutes for human decision-making [11].

Without human input, fully automated network designs are not very accurate.

We discovered that you can't codify every network design guideline or cost/benefit analysis in an automatically generated design since Fiber architecture isn't an exact science. Humans can tell which rules in a design to change or break as well as when to add data to improve the output of an automated design.

More users are happy when all team members can provide data to a finalized network design.

- Planners, for example, are in charge of network architecture and limitations.
- Feedback from the design and the field is provided by designers and field teams.
- Designers can incorporate geographical data during and after ride-out.
- Collaboration with leadership and other partners, as well as the uploading and editing of additional data as needed.

The most crucial aspect is that everyone should have an access to the data and results. Humans control every part of a project with this method, and each network is built from start to finish. Figure 7.3 illustrates the methodical processes of digital domain of optical network.

7.5 INTELLIGENT NETWORK OPERATION AND MAINTENANCE

Intelligent Network Operation and Maintenance allows rapid network forecast, service commencement, and provisioning during the network deployment phase. Traffic forecasting, troubleshooting, and malfunction warning are all included

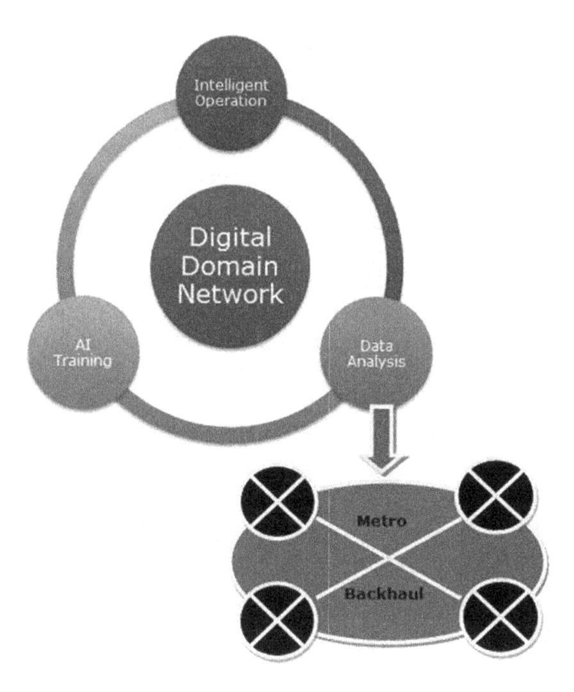

FIGURE 7.3 Digital domain of optical network.

in the operation phase. Adaptive resource management and network slicing are used in the optimization phase to segment the network so that a given service is suitably supported by a network slice in a resource-efficient manner. Artificial Intelligence (AI) in the network cloud engine (NCE) improves O&M efficiency and paves the road for zero-touch optical networks by identifying faults on optical networks and by better anticipating future network resource requirements. To provide the aforementioned functionalities, the network controller must be aware of the critical parameters of the optical network's underlying physical layer [12]. Figure 7.4 illustrates the methodical process of Intelligent Network Operation and Maintenance.

So, how did we make Fiber network design simpler and faster while maintaining high quality?

Auto design, or automated Fiber optic network design, uses software that converts huge volumes of available input data into swiftly created designs, such as GIS, network architecture techniques, and field data. We can lay out networks according to any rules we can establish for cable type, path length, design preferences, and other inputs using our patented algorithms by concentrating the automation in the hands of a capable team of engineers. More users are happy when all team members can provide data to a finalized network design. Figure 7.5 illustrates the Fiber network design aspects [13–14].

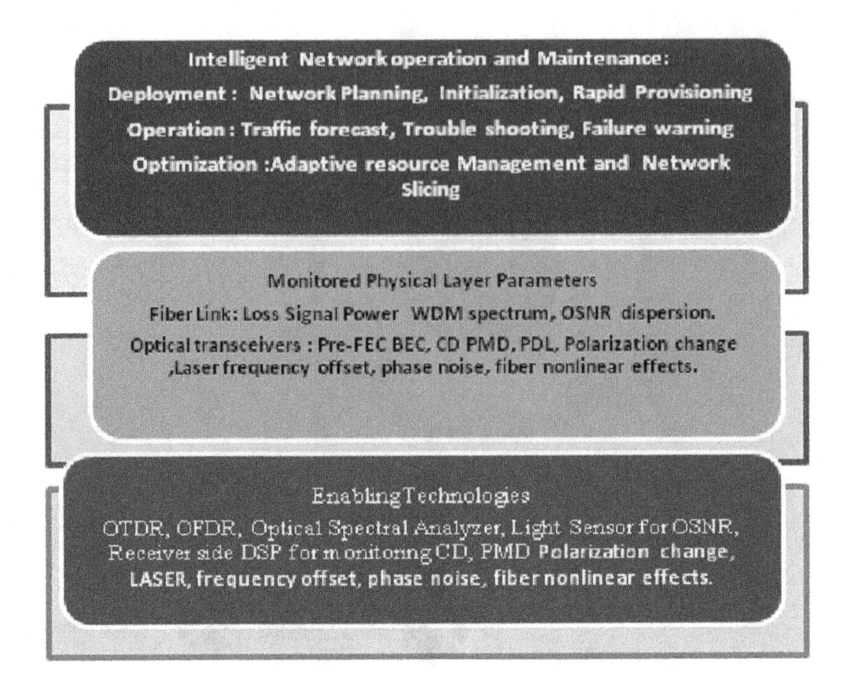

FIGURE 7.4 Intelligent network operation and maintenance process.

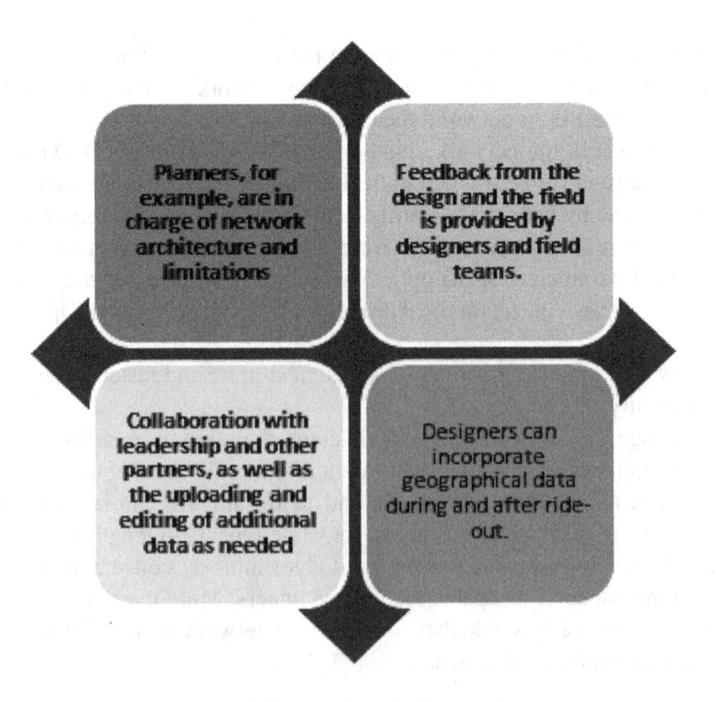

FIGURE 7.5 Simpler and faster Fiber network design.

7.6 APPLICATION NOTES FOR SHOWCASING DIFFERENT NETWORK DESIGN CONCEPTS

Application notes are also helpful for showcasing different network design concepts. The following application notes look at common Ethernet, voice, and data networking systems.

TDM and IP Backhaul Are Used in the Redesign of Analog Radio Systems: A county emergency network on the East Coast was forced to retire four-wire E&M leased circuits by their local carrier.

The principal connectivity for their mission-critical radio system between dispatch and four fire stations was provided by these circuits.

Teleprotection over IP MPLS: The network of an Australian power utility was upgraded to MPLS, but there was no Fiber available to install primary and backup teleprotection links. For 87L Line Differential protection, the JumboSwitch C37.94 IP Gateway passed a range of Latency and Operating Time tests and was implemented as an easy edge solution on their MPLS network.

Legacy and Teleprotection Devices Migrating from TDM to Ethernet: Further power utilities are transitioning from serial communications to Ethernet and TDM systems for simplicity, scalability, and improved interoperability between control rooms and different types of substation IEDs due to the fast convergence of data, voice, and video.

Phones can be Extended Through Ethernet, RS232, or Fiber Optics: It's a very helpful application to be able to change an Ethernet, RS-232, or Fiber Optic connection into a voice network by simply plugging in an analog telephone. It's designed to deliver flexible point-to-point telephone service in even the most extreme environments.

JumboSwitch Voice over IP Application: Two distinct technologies/applications for implementing and integrating voice over a JumboSwitch IP backbone are supported by the TC3848 module.

Smart Grid Applications from JumboSwitch: A smart grid combines electricity distribution methods with current interactive digital technologies to provide intelligent, cost-effective, and reliable electricity delivery from provider to customer.

Applications for JumboSwitch Cellular Backhaul: Mobile backhaul applications have the greatest requirement for increased bandwidth efficiency. Changes in cell phone usage patterns, as well as the introduction of 3G and 4G technology, have resulted in a massive rise in data content for mobility backhaul. As a result, there is a constant demand for additional transmission capacity [15].

Transportation and Intelligent Transportation System (ITS) Applications: In comparison to a "light rail" system, a heavy rail system is an electric train that can accommodate a large volume of traffic. Voice, video, data, and real-time information are all critical applications for heavy train systems.

7.7 CONCLUSION

Design thinking helps not only with the creation of innovative solutions, but also with the identification of the customer's unique problems and the most effective targeting of the customer's requirements. With the rapid advancements in telecommunication and networking technology, there are a number of newly developed techniques that are capable of addressing the subscribers' challenges in specifically in the digital telecommunication era. To solve a problem like connection and update ageing infrastructure, a change in the process and strategy to building new telecommunication networks, the ideas with optimized algorithm and developed fiber optic techniques can be incorporated by bringing out intelligent telecommunication network and management.

REFERENCES

1. Anna Zielińska-Chmielewska, Anna Olszańska, Jerzy Kaźmierczyk, and Elena Vladimirovna Andrianova. *Advantages and constraints of eco-efficiency measures: The case of the polish food industry. Agronomy*, 11(2), 2021
2. www.telecomtoday.in/resources/opinion/application-of-design-thinking-in-strategy-and-innovation.html (Accessed on 14 January 2022)
3. www.businessinnovationbrief.com/design-thinking/telecommunications (Accessed on 16 January 2022)
4. https://biarrinetworks.com/design-thinking-for-telecom-engineering (Accessed on 14 January 2022)
5. www.hcltech.com/blogs/improving-product-support-telecom-software-through-design-thinking (Accessed on 17 January 2022)
6. www.design-thinking-association.org/explore-design-thinking-topics/vertical-markets/design-thinking-in-telecommunications (Accessed on 18 January 2022)
7. https://think.design/industries/telecom-networks-smart-devices-iot/ (Accessed on 15 January 2022)
8. http://citeseerx.ist.psu.edu/viewdoc/download?doi=10.1.1.858.8575&rep=rep1&type=pdf
9. Xiang Liu. Evolution of fiber-optic transmission and networking toward the 5G era. *iScience*, 22, 489–506, 2019
10. S. Varughese, T. Richter, S. Tibuleac, and S. E. Ralph. "Joint Optimization of DAC and ADC Based on Frequency Dependent ENOB Analysis for High Speed Optical Systems". 45th European Conference on Optical Communication (ECOC 2019), 2019
11. www.hcltech.com/blogs/improving-product-support-telecom-software-through-design-thinking (Accessed on 16 January 2022)
12. www.hfsresearch.com/research/design-thinking-telecom-operations (Accessed on 18 January 2022)
13. https://thisisdesignthinking.net/2015/09/ericssons-innova-system-how-to-evoke-employees-entrepreneurial-spirit (Accessed on 14 January 2022)
14. https://uxplanet.org/user-experience-design-the-future-of-telecom-industry-in-the-middle-east-d3cdce937f50 (Accessed on 18 January 2022)
15. http://citeseerx.ist.psu.edu/viewdoc/download?doi=10.1.1.858.8575&rep=rep1&type=pdf

Section III

Applications

8 Design Thinking Applied for the Development of Laptop Workstation to Reduce Postural Risk

R. Mohanraj, M. Senthilkumar, C. Vigneswaran, S. Elangovan, S. Pratheesh Kumar, R. Manish Kumar, R. Naveen Raj, B. Nitish Narayanan and M. Pranaavh
Department of Production Engineering, PSG College of Technology, Coimbatore, India

CONTENTS

8.1 INTRODUCTION

Electronics gadgets such as laptops are widely utilized due to their lightweight and portability features. The laptop has become more popular among students and other software professional which was brought about by technological breakthroughs in memory space and central processing unit speed. The unique sales point of laptops is being portable and that it permits customization in a wide variety of configurations. Laptops are becoming an integral part of life nowadays, and the need for a platform to operate laptops from any place has become a necessity due to the current worldwide situation which directed the shift from the normal office environments. During the usage of laptop, maintaining a correct sitting posture would become cumbersome owing to the numerous musculoskeletal risk factors. Numerous studies reported that placing a laptop on human lap for extensively long

DOI: 10.1201/9781003189923-11

hours would cause harm to the human body. Center of gravity of the body is maintained on its base of support in an ideal working posture, where mechanical stress is minimized by shoulder, elbow, wrist, and interphalangeal joints' movement.

The usage of laptops without maintaining proper suggested posture encourages postural misalignment, which results in biomechanical responses such as spinal arthritis, disc degeneration, and headache. Positioning laptops on a higher platform will tend to hunch the shoulders and placing on lower platforms leads to slouching of shoulders. To reduce the stated ergonomical risk factors, innovative laptop stand has to be designed for the ease of working when it comes to holding and placing it on the platforms. Hough and Nel [1] analyzed postural risks and musculoskeletal discomfort during laptop use among students. The postural risks that students face during the use of laptops at three different positions were studied. Modest postural risk of 69.1%, musculoskeletal discomfort in the form of pain or numbness of 95%, neck stiffness of 47.2%, and neck spasms of 44.4 % were reported based on the survey conducted with students. Gender disparities in adjustments made while using a laptop were revealed by tendencies and statistically significant differences.

Asundi et al. [2] reported on user's postures while working on a computer in their lap and when placed on a desk. The head, neck, and upper extremity postures of 15 individuals were observed using a motion analysis system as they worked on a computer on a desk and on the user's lap. Placing the laptop on the lap increased the downward head tilt by 6° and the wrist extension by 8° as compared to placing the laptop on a desk. Shoulder flexion and ulnar deviation both decreased by 13° and 9°, respectively. In the desk setup, individuals reported reduced discomfort and difficulties. Bubric and Hedge [3] studied the postural risks associated with the use of laptop on a bed. Studies were performed based on the postural risks linked with three common configurations preferred by women while using a laptop on a bed. Although the use of laptop was linked to a number of musculoskeletal risk factors, there is a paucity of research on its use with proper position settings other than a traditional desk and chair. Laptops are used in a variety of configurations because they are designed to be portable, some of which may put the user in danger of damage to their health. Disparities associated in terms of discomfort experienced in laptop usage between males and females were reported.

Chang et al. [4] investigated the locations, furnishings, input devices, and postures connected with laptop usage among a small group of college students. Rapid Upper Limb Assessment (RULA) was performed, and the postural risk factors were detected on the participants while working on their typical workstation setups using digital pictures. It was found that 75% of the participant's preferred standard table and chair arrangement for the laptop usage, while 25% used the laptop in unconventional configurations, such as placing on their laps or in bed. The most common postural risk factors detected were excessive shoulder flexion of 61% and neck flexion of 35%. The results of the RULA test indicated that more postural research was needed. Arborelius et al. [5] reported that instrument does not require any specific assessment of the stress imposed due to the postures of the neck, trunk, and upper limbs maintained during laptop usage. Coding system was used to generate an action list which indicated the level of involvement required to mitigate the injury risk due to physical loading. Moffet et al. [6] evaluated the impact of laptop design and working situation on neck and upper limb posture. Only a minor difference in posture was observed while using the laptop at different situations. The results showed that physical exposure variables were influenced by workstation setup.

The findings reported by researchers show that an extended usage of laptops with an improper sitting posture leads to misalignment of the spinal cord eventually leading to back pains particularly while resting on lap [7–8]. To alleviate these issues, it's highly important to ergonomically design a laptop workstation which should be light in weight, portable, and fitting in a backpack along with the laptop. Laptop workstation should be lightweight, and also it should be capable of withstanding the load generated by the user's hand while using the laptop. Design thinking methodology provides a solution-based approach to address complex problems. Design thinking strategies such as Empathize, Define, Ideate, Prototype, and Test have been adopted to develop a laptop workstation. Considering the ergonomic factors, laptops workstation is designed with highly appreciable features such as flexibility in all environments and is easily deployable. The proposed design would be compatible for users of all age groups with respect to the aspects of varying height and inclination. The workstation designed to avoid risks posed by them makes an alignment of posture avoiding musculoskeletal problems and enables the usage of the laptop at a comfortable posture.

8.2 DESIGN OF LAPTOP WORKSTATION

A laptop workstation has been designed considering the ergonomic factors empathized by individual users in a society. Based on the preliminary survey reports, the purpose of the product has been defined, and the product featured with flexible mechanisms was designed in PTC Creo software. The CAD model and various features of the proposed design are shown in Figure 8.1. The design of laptop workstation includes the features such as perforated base plate in order to improve heat dissipation as shown in Figure 8.2 and thigh support to make the platform stable during usage.

Height and inclination adjustment mechanism for 12 different positions was incorporated in order to improve the convenience of the user. The holding components of the laptop station are designed in such a way that they can be clamped to a stable surface in a chair or table.

The design enables the user with maximum comfort, where the slot and latch mechanism allows the users to adjust the distance between the user and the laptop position point as shown in Figure 8.3 and also adjust the angle of the platform for user convenience as shown in Figure 8.4.

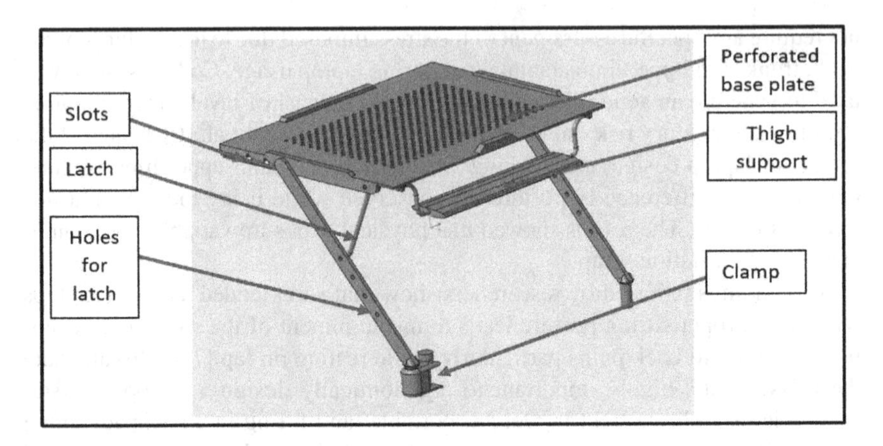

FIGURE 8.1 Proposed design of laptop workstation setup.

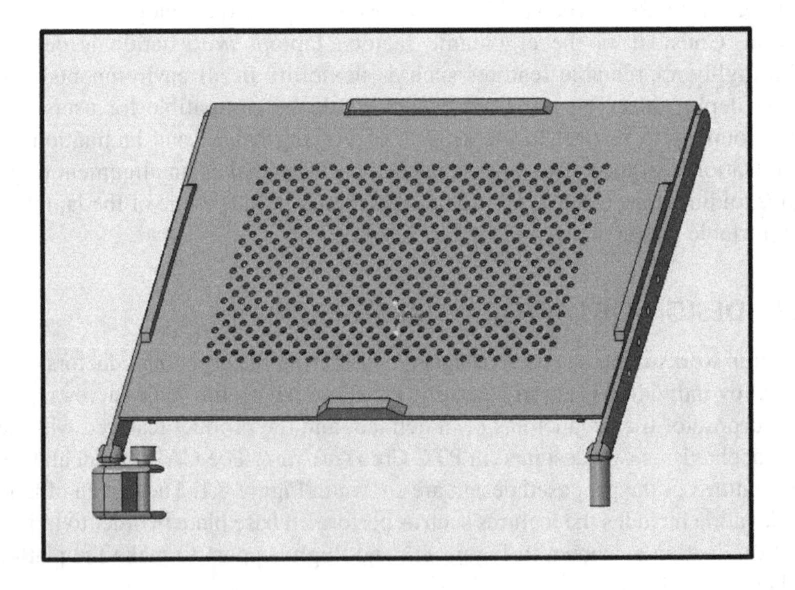

FIGURE 8.2 Perforated base plate for heat dissipation.

The designed clamps can be rotated by an angle of 90°, which is a feature that provides a provision of clamping the laptop station in two different orientations. The dimensions of the laptop station are finalized based on the standard, so that it fits in a backpack which can be carried anywhere and deployed to use within a minute of time.

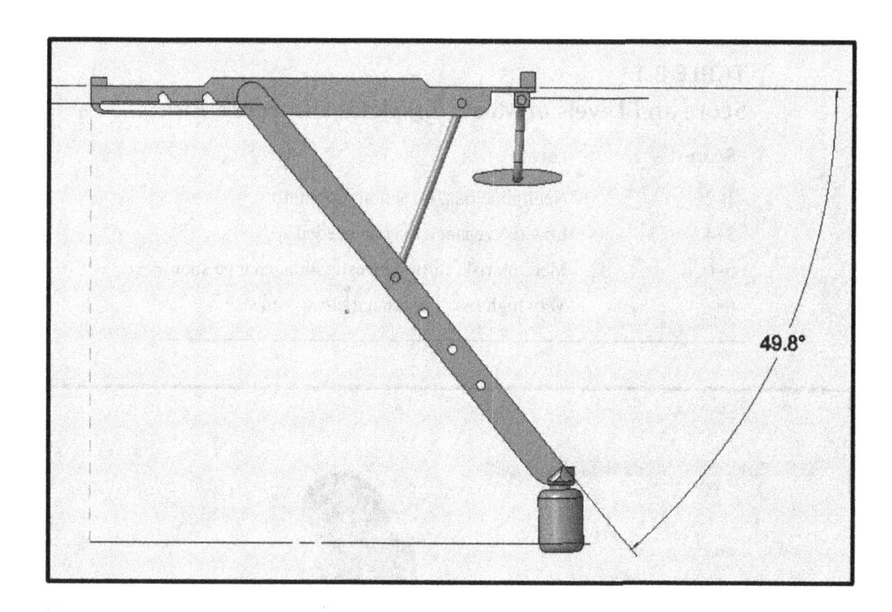

FIGURE 8.3 Slot and latch mechanism for distance adjustments.

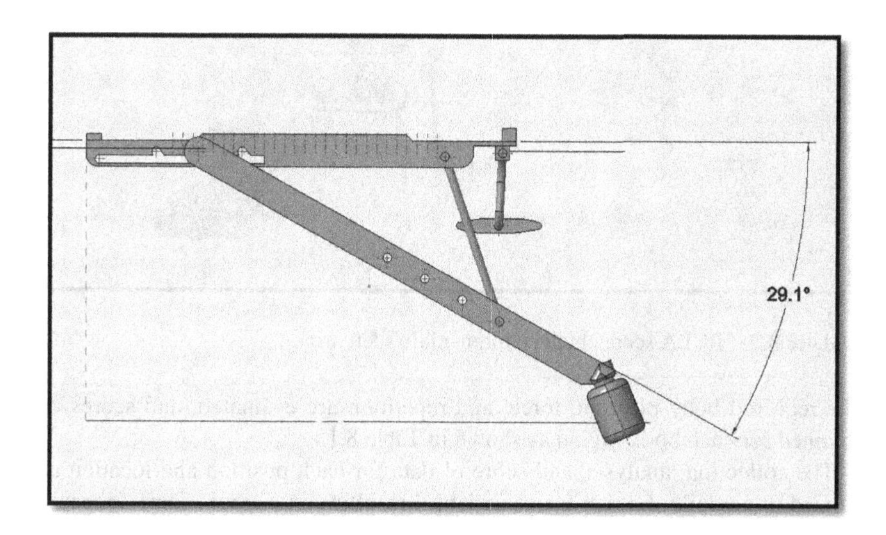

FIGURE 8.4 Provision for angular adjustments.

8.3 DESIGN ANALYSIS USING RULA

Rapid Upper Limb Assessment (RULA) tool is utilized to assess the exposure of individuals to ergonomic risk factors linked to upper extremity musculoskeletal disorders. The biomechanical and postural load on the neck, trunk, and upper extremities during working are considered for the RULA ergonomic assessment.

TABLE 8.1
Score and Levels of Musculoskeletal Disorders' Risk

Score	Level
1–2	Negligible risk, no action required
3–4	Low risk, change may be needed
5–6	Medium risk, further investigation, change soon
6+	Very high risk, implement change now

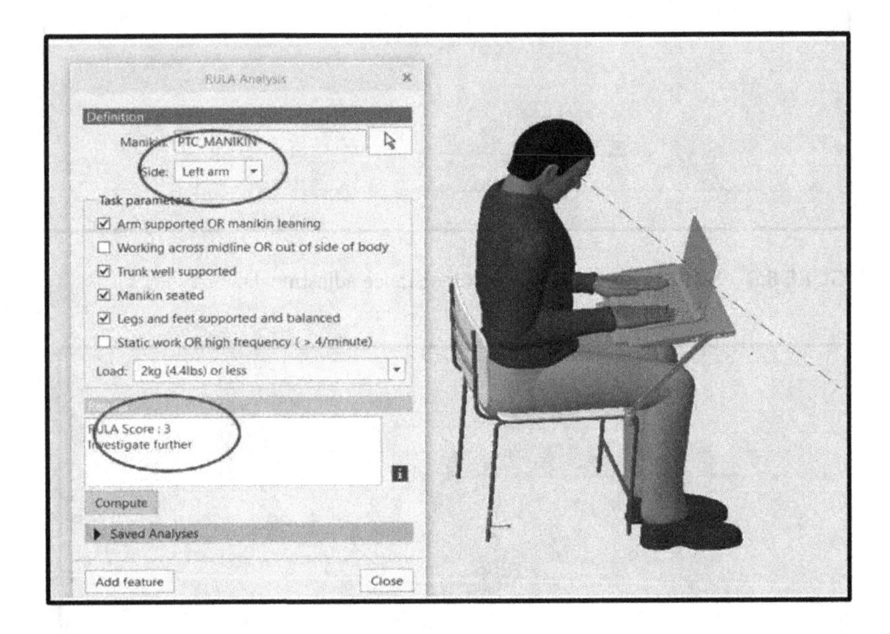

FIGURE 8.5 RULA score obtained for manikin's left arm.

The required body position, force, and repetition are evaluated, and scores are obtained for each body region as shown in Table 8.1.

The collection, analysis, and score of data for each position and location are utilized to compile the risk factor variables resulting in a single score for representing the level of musculoskeletal disorders' risk.

The laptop workstation was clamped to a chair, and a manikin was placed on the chair with a vision fixed onto the laptop screen. Using the reach option, the hands were placed on the keyboard to represent the typing action. Scores for the left and right arms of the manikin were obtained as 3 as shown in Figures 8.5 and 8.6.

The risk factors for three different working postures were examined. Under the working posture, static posture analysis with arms supported and the person leaning criteria were selected for analysis. A score of 3 was obtained as an output from the RULA tool for posture 1 as shown in Figure 8.7, where all the regions

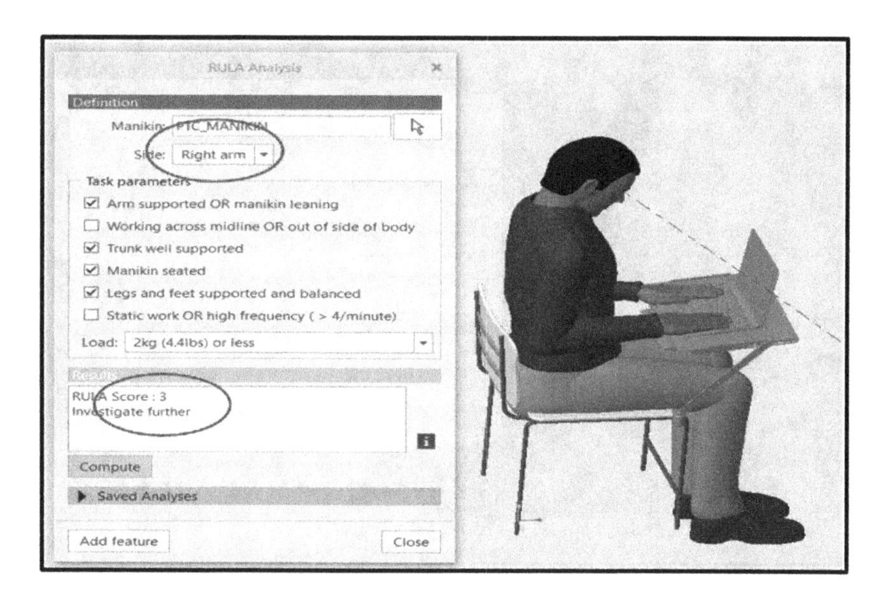

FIGURE 8.6 RULA score obtained for manikin's right arm.

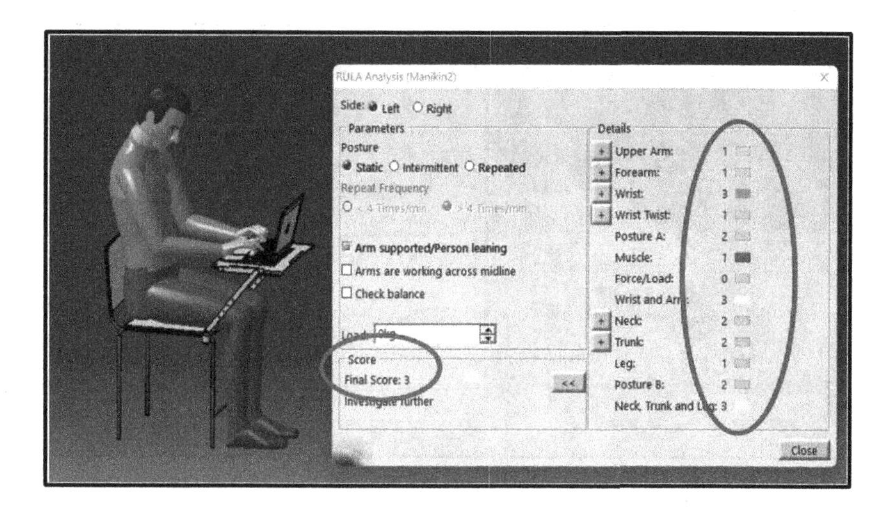

FIGURE 8.7 RULA score obtained for posture 1.

of the body scored less than or equal to 3, making the position preferable for an average-sized person.

For posture 2, the height of the workstation was adjusted to minimum and analyzed. This posture scored 3 for a manikin of reduced dimension, making it preferable for a shorter person as shown in Figure 8.8. In the case of an average-sized person, the spine tends to bend forward, leading to a slouching posture and making it difficult to work for long hours.

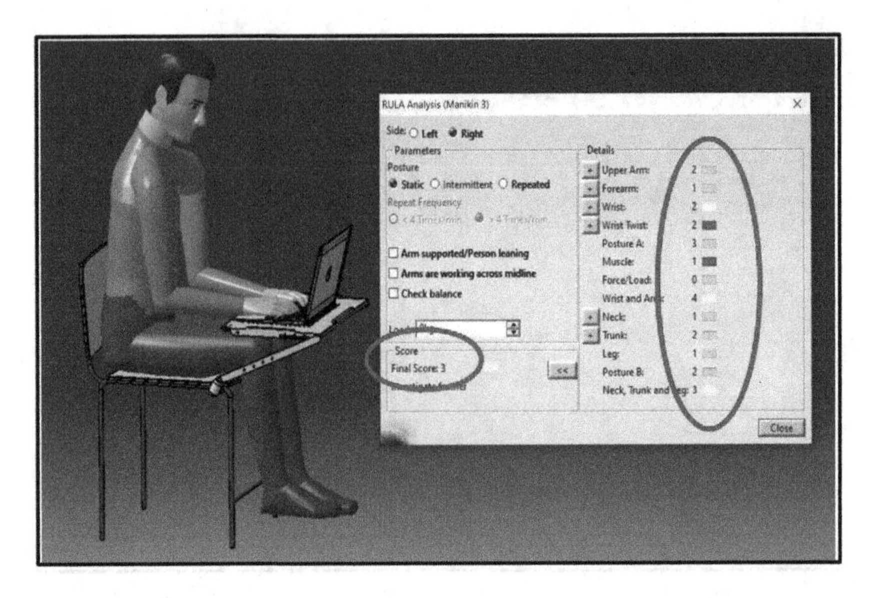

FIGURE 8.8 RULA score obtained for posture 2.

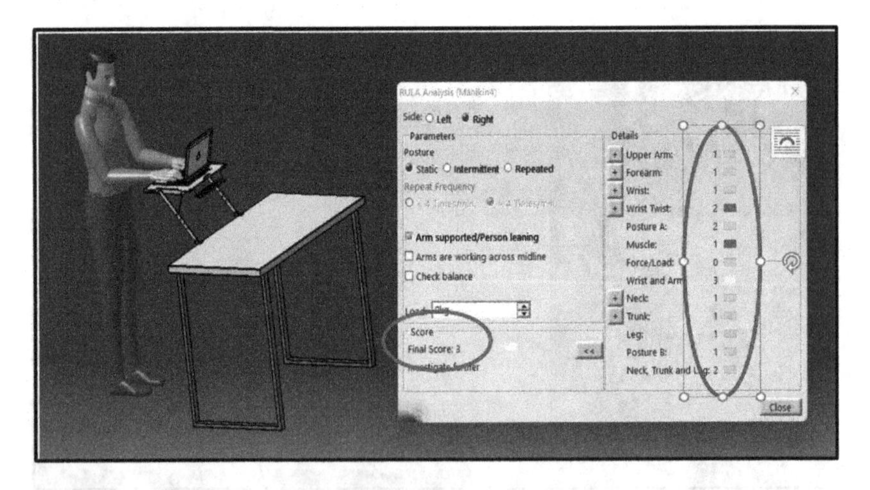

FIGURE 8.9 RULA score obtained for posture 3.

Standing position was considered for the analysis of posture 3 as shown in Figure 8.9, where the laptop workstation is clamped to a table instead of a chair. The score 3 was obtained from the RULA tool which indicates that the position is preferred for an average-sized person.

8.4 FABRICATION OF DESIGNED LAPTOP WORKSTATION

The designed model of laptop station was fabricated, and the feedback about the product had been received from the different individuals. Individual components were fabricated with reference to the design finalized. After fabrication, the individual components of the prototype were assembled and fastened together as shown in Figures 8.10 and 8.11.

The assembled product has been tested by analyzing the functions and operations of the individual components. As per the design, the attainment of the angular position of the individual specific components has been verified and found to be perfect for the real applications.

The product fabricated was subjected to real application scenario and analyzed for the load-carrying capacity. For analysis, the prototype was fixed to the table as a user interface with laptop as shown in Figure 8.12. The user felt that physical discomfort has been reduced to great extent.

Then the prototype was clamped to a stool and the product tested with a user working in a sitting posture on the stool as shown in Figure 8.13. The user felt

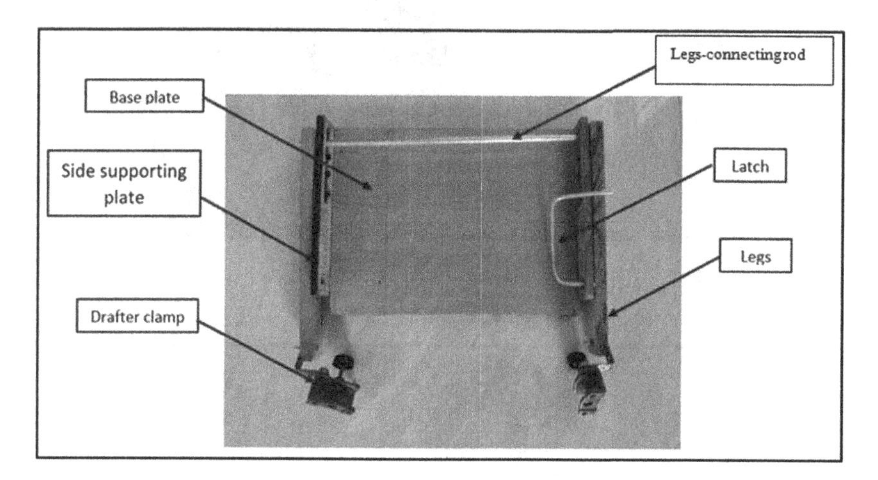

FIGURE 8.10 Components of the laptop workstation.

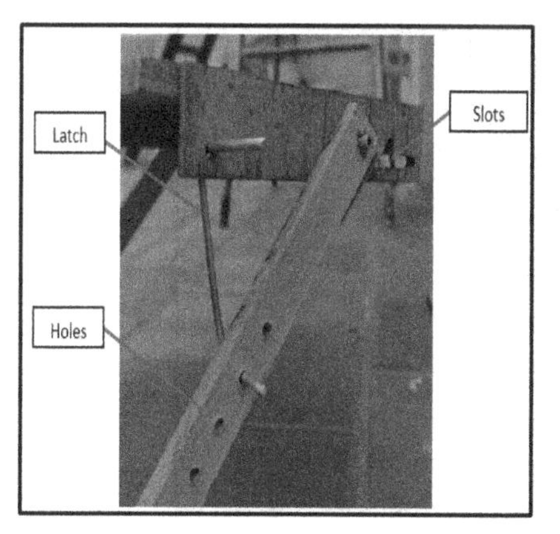

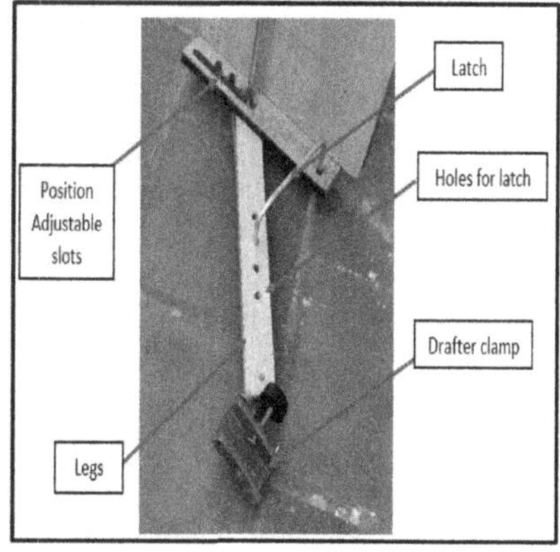

FIGURE 8.11 Assembly of the laptop workstation.

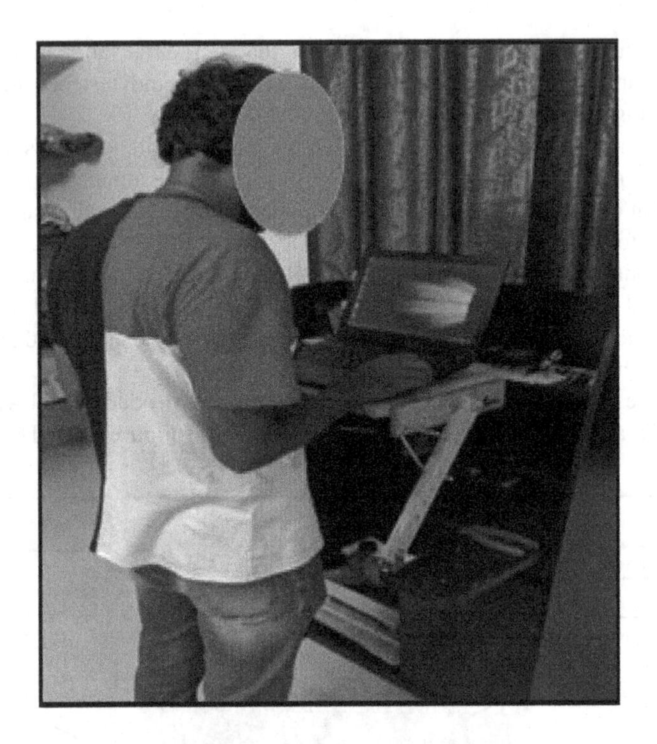

FIGURE 8.12 Laptop workstation fixed on the table.

FIGURE 8.13 Laptop workstation fixed on the stool.

comfortable in using the laptop at sitting posture for long hours and did not report about the physical stress at the joints of the body.

8.5 CONCLUSIONS

The workstation for laptop has been designed and fabricated based on the ergonomic factors. The results obtained from the RULA revealed that the product is suitable for real applications. It has been identified that the usage of the product reduced the stress and fatigue on the human body caused during working. Therefore, the proposed design of the product can be used comfortably for long hours. Design thinking strategies have been successfully adopted for the development of laptop workstation.

REFERENCES

[1] Hough, PA and Nel M. (2017) Postural Risks and Musculoskeletal Discomfort of Three Preferred Positions During Laptop Use Amongst Students, *South African Journal of Occupational Therapy*, 47:1, 3–8.

[2] Asundi, K, Odell, D, Luce, A and Jack TD (2010) Notebook Computer Use on a Desk, Lap and Lap Support: Effects on Posture, Performance and Comfort, *Ergonomics*, 53:1, 74–82.

[3] Bubric, K and Hedge, A (2014) Postural Risks Associated with Laptop Use on a Bed, *Proceedings of the Human Factors and Ergonomics Society Annual Meeting*, 58:1, 586–590.

[4] Chang, C, Amick, BC, Menendez, CC, Robertson, M, Pino, RJ and Dennerlein, JT (2008) Where and How College Students Use Their Laptop Computers, *Proceedings of the Human Factors and Ergonomics Society Annual Meeting*, 52:15, 1010–1014.

[5] Arborelius, P, Wretenberg, P and Lindberg F (2007) The Effects of Armrests and High Seat Heights on Lower-Limb Joint Load and Muscular Activity During Sitting and Rising, *Ergonomics*, 35:11, 1377–1391.

[6] Moffet, H, Hagberg. M, Hansson, RE and Karlqvist, L (2002) Influence of Laptop Computer Design and Working Position on Physical Exposure Variables, *Clinical Biomechanics*, 17:2, 368–375.

[7] Gurusharan Dand Patnaik, MMM (2016) Alternative Design of C Clamp for Minimization of Counter Weight Using FEM, *International Journal for Ignited Minds*, 3:8, 98–101.

[8] Davis, GK, Susan, EK, Denise, D, Gerding, T, Naylor, J and Syck, M (2020) The Home Office: Ergonomic Lessons From the New Normal, *Ergonomics in Design: The Quarterly of Human Factors Applications,* 1:1, 1–7.

9 Using Design Thinking to Develop a New Product

Case Study of Developing a Novel Medical Device

Blaž Zupan[1], Vladimir Pakrac[2], and Anja Svetina Nabergoj[1]
[1]University of Ljubljana, Ljubljana, Slovenia
[2]MESI Ltd, Ljubljana, Slovenia

CONTENTS

DOI: 10.1201/9781003189923-12

9.1 INTRODUCTION

As consumers have become more informed and accustomed to rapid innovation—they expect Uber, Netflix, and Amazon-like level service in all aspects of their lives (Nilsson & Sheppard, 2018), leaving little room for having to use inferior products. Innovation is increasingly being seen as the most important way to differentiate and maintain competitive advantage in the twenty-first century. In certain industries, like healthcare, continuous improvement of products and services is governed by regulations and guidelines enforced by specific industry standards. As Henrik Norström, the deputy CEO of Brighter AB, a global mobile health tech company dedicated to innovative designs on FDA guidelines stated, "We have to follow 36 standards, sometimes more" (Nilsson & Sheppard, 2018).

In healthcare, fostering user-centricity has been an important effort among organizations around the world, proving that switching from a focus on disease to a patient-centered approach can produce improved patient outcomes, as well as hospital outcomes (Nilsson & Sheppard, 2018). There is no doubt that design can help craft exceptional patient experience. In this chapter, we will explore how a medical device company developed internal capabilities for user-centered innovation, developing new products that are both desired by users as well as are up-to-date with the regulations and guidelines. They implemented design thinking, which is an innovation methodology that puts the user and the user experience at the heart of product or service development—leading to both incremental improvements of existing products and services, as well as to the development of entirely new products or services.

Design thinking is usually defined as a process, set of mindsets, and a set of tools. As a set of mindsets, design thinking is characterized by several fundamental principles: a strong focus on the explicit and underlying needs of customers, rapid prototyping, iterative approach to user testing, and introducing diverse perspectives through cross-functional collaboration. As a process, design thinking is defined as a user-centric innovation process that consists of five stages: empathize, define, ideate, prototype, and test. As a tool, design thinking refers to the application of numerous methods and techniques from various disciplines: design, as well as engineering, computer science, anthropology, and psychology. Today, an increasing number of companies, consulting firms, and universities are using design thinking, constantly expanding and rethinking its meaning (Brenner et al., 2016). The aim of this process is to put the patient at the center, to gain better insight and understanding of user needs, and to develop products or services that are deeply grounded in these needs.

Design thinking promotes the desire to understand holistic patient experiences from morning until night. This means understanding the full patient experience, and not just the direct symptoms. It means understanding the pain and frustration that diseases bring, instead of narrowly focusing on devices or drugs. It uses specific activities of ethnographic and anthropological nature in the research phase, which can include empathy interviews, immersion or putting oneself in a user's shoes, shadowing or living with users, and other ethnographic research approaches (Hammersley, 2016). The aim is not just to identify explicit user needs, but more importantly to uncover underlying needs that have not yet been recognized. As such, these represent an untapped potential for innovation and can increase company competitiveness, as it aims to develop novel solutions to addresses such needs.

This way of working, unlike traditional market research, offers a different and, above all, more comprehensive view of both market and user requirements. In the healthcare industry, this means involving key stakeholder perspectives including those of patients, doctors, nurses, and other healthcare professionals as well as patient's family members. These offer deeper insight into the user experience and life cycle of the product—putting user needs at the forefront. As Sebastian Liedtke, the head of digital-design transformation at Roche Diabetes Care stated, "Getting closer to patients means conducting a full range of ethnographic research using techniques such as focus groups, home visits, and family interviews" (Nilsson & Sheppard, 2018). With such approach to innovation, companies can add real value to the product or service and often redefine its use.

The aim of this chapter is to present how a medical device company developed an internal capability for user-centered innovation, and how they leveraged design thinking abilities and mindsets throughout the development of a novel medical device.

9.2 DESIGN THINKING

Martin (2009) defines design thinking as the ability to create better solutions than those that already exist. He illustrates this idea with a choice between solutions A and B, neither of which provides an optimal solution to the problem. He says that in such a situation, the person using design thinking chooses neither solution but is able to create his or her own solution, C, which is better than the original solution.

9.2.1 Historical Development of Design Thinking

Design thinking has emerged from the study of theory and practice in several disciplines and sciences—as an approach to solve the human, scientific, technological, and strategic innovation needs of our time. In Figure 9.1, the timeline of the formation of design thinking is illustrated.

In the mid-1960s, Horst Rittel, a Design Theorist known for inventing the term "Wicked Problems" (extremely complex problems), wrote about problem-solving in design. Wicked problems are at the very heart of design thinking, because these complex problems require a collaborative methodology that involves a deep understanding of humans (Dam & Siang, 2018).

In the sciences, the notion of design as a "way of thinking" is credited to computer scientist and Nobel Prize laureate Herbert A. Simon's 1969 book, *Sciences of the Artificial*, where he contributed many ideas that are now regarded as postulates of design thinking. He spoke of rapid prototyping and testing through

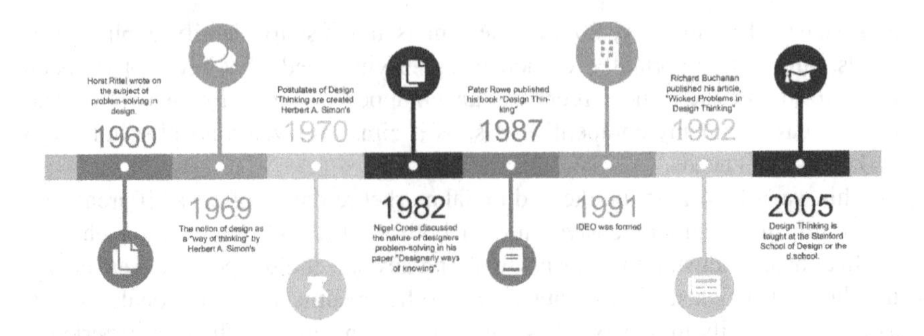

FIGURE 9.1 Timeline of design thinking formation (adapted from Liakhavets & Petukh 2019, with permission).

observation and concepts which form the core of many design and entrepreneurial processes today (Dam & Siang, 2018). In 1982, Nigel Cross discussed the nature of problem-solving for designers in his paper, "Designerly ways of knowing" (Dam & Siang, 2018). After that, Peter Rowe, Director of Urban Design Programs at Harvard published his book, *Design Thinking*, in 1987, concentrating on the way that architectural designers approach their tasks through the lens of inquiry (Dam & Siang, 2018).

The evolution of the innovation approach continued with its popularity in scientific journals and professional design agencies in the 1990s. In 1991, IDEO was formed with the merger of design agencies David Kelley Design and ID Two. The latter initially focused on providing conventional industrial and graphic design services. In early 2001, the company experienced a significant increase in demand for solutions to broader problems. For example, a healthcare organization asked for help in restructuring its entire organization, an established manufacturing company accepted a brief to better understand its customers, and a university commissioned IDEO to redesign its teaching processes. For IDEO, this meant a shift in thinking from designing tangible entities to designing user experiences. David Kelley, co-founder of IDEO and later co-founder of the Hasso Plattner Institute at Stanford University, said that back then, when asked what designers did, he always included the word "thinking" in his answer. Thinking, he said, is the ethos of a designer's work. From that point on, the term design thinking was used, referencing the methodology of holistic problem finding and problem solving. IDEO later popularized the methodology worldwide, and it is now a leading organization in product and service design (Brown & Wyatt, 2010).

Design thinking is not about design. It is about helping companies and individuals to think differently about strategic options and system impact. "Design thinking is about applying the methodology and approaches of design and designers to a broader set of issues and problems in business and society" (Brown, 2008). It is an essential tool for simplifying and humanizing, as Brown said, "thinking like a designer can transform the way you develop products, services, processes—and even strategy" (Brown, 2008).

9.2.2 The Design Thinking Process

There are various models of design thinking based on different aspects of design situations drawing on the foundations from psychology, anthropology, design methodology, etc. (Dorst, 2011). The number of phases varies between different theoretical models, and the differences are mainly because some models combine several phases into one or vice versa (Zupan & Svetina Nabergoj, 2014). Figure 9.2 overviews the process breakdown.

In this study, we will follow Hasso Plattner Institute of Design at Stanford University, which teaches design thinking as a five-stage process as seen in Figure 9.3.

9.2.3 Empathize

The empathize phase is the first phase of the design thinking process and helps the development team to identify and understand its users in the context of their specific task, how they use things, and why they do things the way they do. Understanding the user is regarded as a key phase because it helps designers gain insight into the explicit and implicit needs of the users (Kolko, 2014). To best understand the users, designers use different ethnographic techniques and collect data by engaging with users and through observation (Lojacono & Zaccai, 2004; Leonard & Rayport, 1997). The aim of these techniques is to get as close as possible to the users' view of the problem, which is a prerequisite for understanding their experience and getting to know their real wants and needs. This phase can also include consulting with field experts who interact with users as part of their work and have an insight into the experiences, values, and beliefs of the people we are designing for (Interaction Design, 2020).

9.2.4 Define

The second phase of design thinking is the "define" phase. The aim of this phase is to structure the information that the design team has received during the empathy phase. Defining the problem is key to forming a development perspective for that problem, and all further developments are dependent upon this. During this phase, the development team combines user insights to identify the problem(s) to be addressed (Derenda Mujezinović, 2017). This enables focus and embraces a holistic theme—it inspires the entire development team, prompts an evaluation of the concreteness of ideas, enables the team to make independent decisions, provides a space for generating ideas based on the question "How could we?", takes into account the understanding and mindset of the users addressed, protects against the impossible notion of developing a product to satisfy all users, forces review and possible change through the principle of experiential learning, and ultimately leads to the pursuit of innovation.

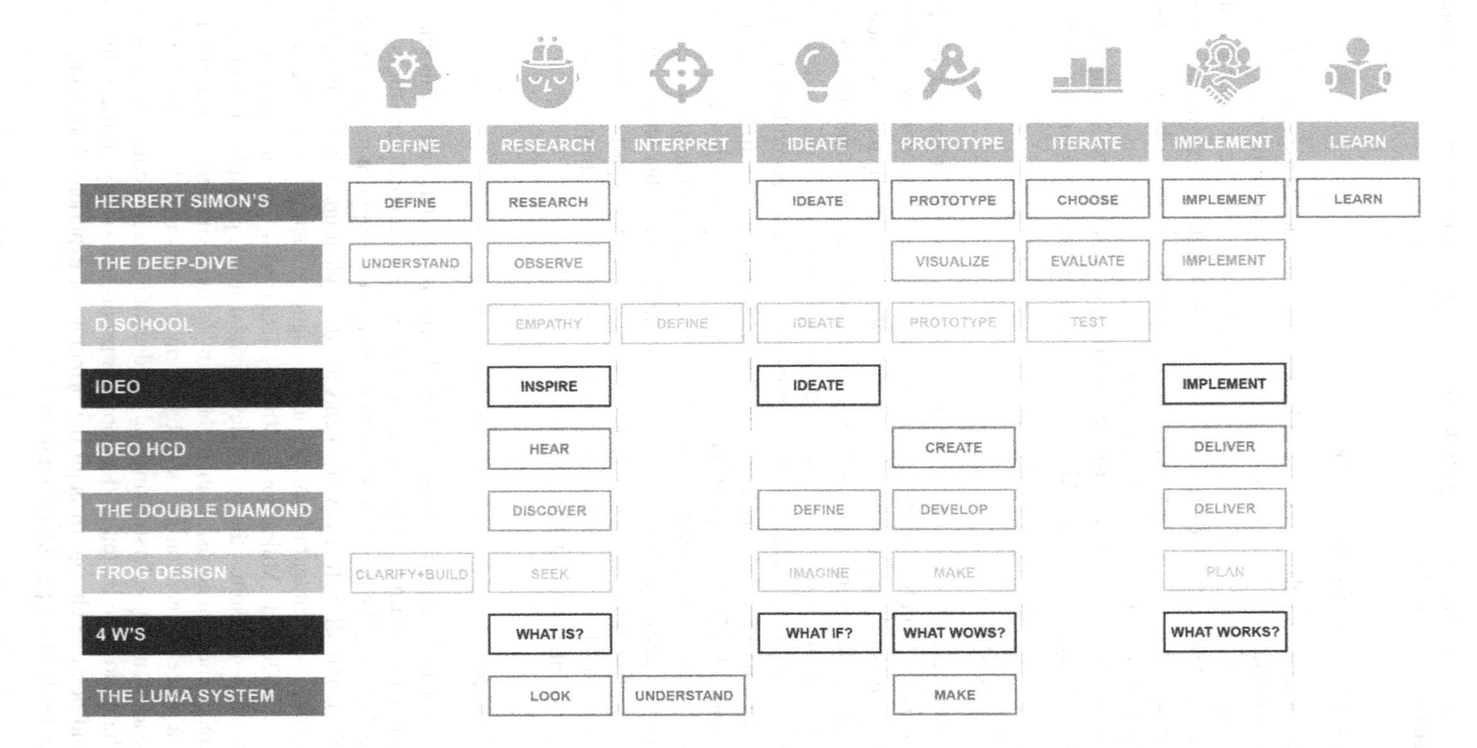

FIGURE 9.2 The design thinking process overview (based on Simon, 1988; Dam & Siang, 2018; Brown & Wyatt, 2010; Design Council, 2007; Liedtka & Ogilvie, 2011).

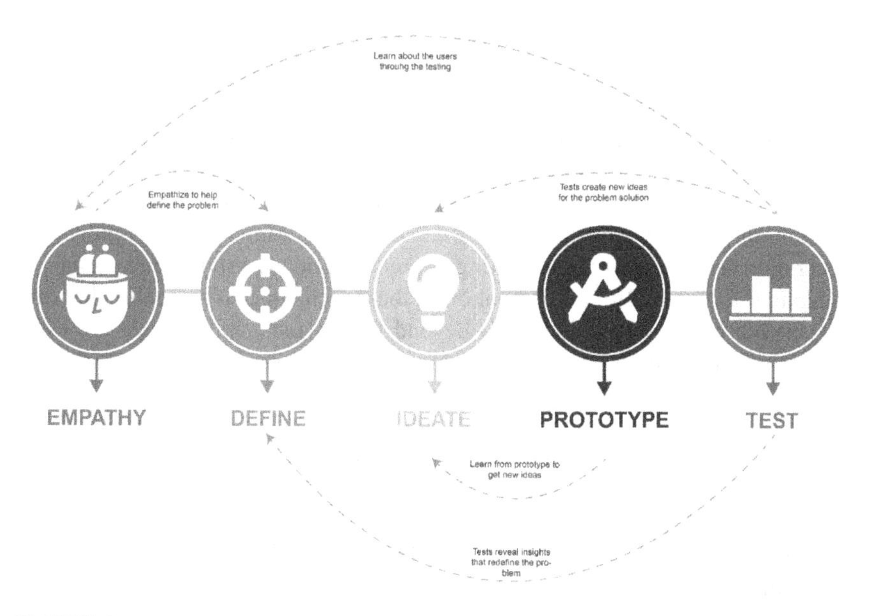

FIGURE 9.3 The design thinking processes (design by Liakhavets & Petukh, 2019; based on Dam & Siang, 2018).

Fry (2009) identified three key questions that a development team should ask when defining a problem:

1. What is the product or service about?
2. Why is the product or service being used?
3. How do we use the product or service?

These questions force the development team to consider why they are developing the solution in the first place. So, for a company, this stage manifests a business opportunity, crucial for developing further ideas about future products and functions. When defining the problem, the development team synthesizes the information from empathy work by focusing on what stood out during the user research, what was most surprising, or whether there were interesting tensions or contradictions emerging from the user stories.

9.2.5 Idea Generation and Selection

Idea generation is the transition from an identified problem to a tangible solution. Therefore, the idea generation phase is an opportunity for the development team to combine their understanding of the user and their needs, with their ability for creative problem-solving that will lead to the creation of physical products.

The idea generation phase is about generating a high volume of ideas for how to solve the problem that was defined in the previous phase, considering a wide range of possible solutions and selecting one that will spawn further development.

In order to generate ideas, design team combines individual and team techniques in order to develop many diverse ideas. One of the most popular techniques is brainstorming (Gregersen, 2018), which is fast paced and fun technique for building on ideas by using "yes, and . . ." technique. Among other popular idea generation techniques are brainwriting, group ideation workshops, and six hats of thinking (De Bono, 2017). The goal of all these techniques is to enable the team to generate many diverse ideas, go beyond the obvious, and remove the usual constraints that limit the thinking only to what is possible. It is important that the team accepts and promotes broad, creative thinking during this phase by creating a safe environment to suggest any idea that comes to mind and by employing techniques that help defer criticizing and judgement of ideas (Lucas & Nordgren, 2020; Sutton & Hargadon, 1996).

Ideation is one of the phases of design thinking process, which is probably repeated the most during the new product development. Ideation is used whenever the team needs to generate multiple options to choose from. For example when the design team moves back from the prototyping phase with a list of things to be improved or a list of bugs to be resolved, it enters an ideation phase in order to discover ways to solve these challenges.

Another critical aspect of ideation is idea selection. This is where the design team needs to transition from divergent to convergent thinking and select the ideas that they will carry further into the development. There are many different selection techniques including selection criteria and plotting ideas in 2X2s. The most important aspect of idea selection is that it needs to be based on the potential of each particular idea and not on feasibility alone. This is how the team can retain the innovation potential and carry it into the next phase of the process where they can develop prototypes.

9.2.6 Prototyping

Once the team has come to a consensus on the idea they want to develop further, they need to bring it to life by building a prototype that can be tested with users—identifying what works and what doesn't—before finally implemented it. Prototypes take different shapes and forms, which can mean a sketch on a piece of paper, a set of sticky notes with a storyline, or actual physical objects like for example 3D printed part that looks quite similar to a final product.

The main objective at this stage is to allow the development team to test the key assumptions that were the basis for their idea and test them cheaply and quickly with end users before committing to the final solution. The main benefit of early and low-resolution testing is that it is significantly less expensive than final product testing (Dow et al., 2010). In the example of the 3D-printed object, producing it is significantly less time consuming and expensive than making a mass-produced plastic piece. This means that the team can always refine, change, or revise their idea based on user feedback, and as such prototyping allows for a quick and inexpensive start (Martin, 2014; Sutton, 2007). A prototype is anything that can help the user better visualize an idea and imagine the solution as if it already existed. Prototyping happens several times throughout the product development process, and the more mature the solution is, the more refined the prototypes become.

9.2.7 Testing and Implementation

Testing is the phase where the development team or team member responsible for testing solicits feedback from potential end users. Testing is a learning process that allows the team to refine the prototype. The purpose is for the team to refine, correct, or modify the prototype based on lessons learned. Through user testing and feedback, the development team often returns to earlier phases of user understanding and ideation based on the transition between phases described earlier.

According to the Hasso Plattner Institute of Design model, the testing phase is the final phase of design thinking, closely intertwined with the prototyping phase; unsurprisingly, these two phases are often combined into one. However, despite the fact that it seems like a final stage, we want to emphasize the importance of iteration, which means that after testing the team returns to earlier phases of the process, based on what they learned during testing.

A prototype serves no purpose unless the team can test it, try it out, and either confirm or disprove conceptual assumptions based on those tests. The prototype should be tested by both test team and end users. Only this way can the design team can determine what is poorly done, what might be missing, and, ultimately, what needs to be changed.

Ideally, testing is conducted in an environment where the product will actually be used—providing users the opportunity to pick up the prototype and interact with it, as if the product already existed. Testing yields the best results when users are left to interpret how products are meant to be used, and the designing team is taking detailed notes or recording the session in order to be able to avoid most common biases in user research (Pinder, 2022). This way, the team can monitor and record reactions that would otherwise be overlooked. It is important for the team to focus not only on what users are doing right, but also on all the things users do wrong, or fail to do, despite expectations.

9.3 USING DESIGN THINKING IN PRACTICE

Seidel and Fixson (2013) state that the use of design thinking can be seen in companies and organizations seeking innovative responses to rapidly changing market and competitive demands. Here, the innovation approach serves as a tool for designers and other employees to address business challenges. Given the adaptive nature of the application of design thinking as described before—business leaders typically do not have a problem with where to apply the method, but rather with how and how intensely to apply and integrate it.

Despite evidence that the design thinking process is typically present in the implementation of innovative ideas, the introduction of the model at the organizational level requires strategic adjustments. In fact, the foundation for the successful integration of the innovation approach within an organization lies in management awareness—in their recognition that the use of a design approach can provide an array of benefits. A survey by the British Design Council (Design Council, 2007) found that in the top half of the 250 most innovative companies in the UK, one senior manager is always a designer. Design thinking can therefore only become a source of competitive advantage if it is integrated into a company's culture, processes, systems, and other structures (Rosenberg et al., 2015). According to Liedtka (Mural, 2018), in applying the innovation approach of design thinking, organizations seek:

- better ideas,
- decreased risk and investment,
- increased quantity of ideas, and
- increased flexibility within the organization.

Although there is no clear answer when it comes to how many organizations use design thinking, Schmiedgen and colleagues (Schmiedgen et al., 2015) state that it is present in a wide range of organizations across virtually all industries. Dunne (2018) notes that design thinking offers organizations in both the public and private sectors an opportunity to develop unique products and services that address otherwise hidden user needs. Organization leaders have adopted design thinking as being either an innovative source of competitive advantage, or to innovate and manage organizational change in a world of rapidly changing customer demands, supply chains, and hyper-competition. Martin (2009) sees design thinking as the "next big thing" in creating competitive advantage, while Dunne (2018) compares design thinking to a vehicle that guides an organization through organizational change and development.

Any product or service development needs at least basic input requirements. These are usually written down in a document that forms the basis for product development. The requirements may change during development, depending on user testing results of intermediate product iterations. The latter can take many forms at varying stages of development. They can be tested as sketches, models glued together, 3D printed parts, etc. A prototype can be anything that gives the end user a better idea of what the final product will look like (Houde & Hill, 1997; Buchenau & Fulton Suri, 2000). Prototyping provides development teams testing tools to present to users and therefore plays one of the most important roles in design—placing users at the centre of development. However, the function of prototyping goes beyond testing. Prototypes are key to fostering communication between developers and users, bridging different actors involved in the development process (Dunne, 2018).

Different types of risks in new product development need to be addressed with different levels of care. Notable examples of industries with input requirements, which carry a higher degree of risk, include both automotive and medical devices. Aside from the fact that the input requirements for developing a new car are incomparably higher than the input requirements for developing a new mouse or, a citrus juicer per se, many of these products also carry high risk. In such cases, design thinking strongly advocates working in interdisciplinary teams. Interdisciplinarity in a team also encourages a higher level of convergent and divergent thinking (Chasanidou et al., 2014). Consequently, in companies and organizations that work with so many input requirements and risks, diverse teams consisting of engineers, designers, product, and project managers are the norm. These diverse teams need to collaborate, sharing their experiences and specific skill sets, ultimately leading to the development of the highest quality products possible. The nature of engineers and designers creates a unique challenge, as each view goals through a completely different prism. For example a designer may place more emphasis on feel and aesthetics when designing the shape of a car, but less on safety and environmental impact. On the other hand, a mechanical engineer is increasingly focused on lowering manufacturing costs and keeping production simple. Given these differences among and within development teams, product managers have become a popular solution to ensure that everyone involved works together in harmony. Assuming that designers have the knowledge to design an aesthetically perfect product and engineers have the knowledge to manufacture said product, it is the product manager's job to ensure that the final product meets the input requirements.

9.4 CASE STUDY: USING DT TO DEVELOP A NOVEL MEDICAL DEVICE

Although design thinking is becoming an increasingly recognized and popular innovation approach within organizations, the literature featuring concrete examples of design thinking use in product development is sparse. Research related to design thinking still largely focuses on process descriptions and individual phases of design thinking, rather than on how companies implement design thinking in product and service development.

One industry that has adopted design thinking rather quickly has been healthcare, specifically medical device manufacturers. The reason is that MedTech has been highly traditional, and product development has long followed a traditional waterfall approach. This caused long lead times from idea to market and led to sub-optimal technology-driven solutions. As Sebastian Liedtke, formerly in charge of digital–design transformation at Roche Diabetes Care shared in his interview,

> Historically, many medtechs have been technology-centric rather than user-centric. As a result, they are structured in tech silos: mechanical, electrical, software, and so on. Each of these silos has optimized its own individual part rather than addressing the entire patient journey.
>
> **(Nilsson & Sheppard, 2018)**

In this chapter, we aim to illustrate a practical application of design thinking in a Slovenian medical device company, MESI Medical. We will focus on innovative product development from its early stages and through production. With this research, we aim to provide insight into how the company implemented the design thinking process, as well as how the company fostered design thinking mindsets and developed internal capability for user-centered innovation.

9.4.1 Methodology

In our analysis, we use a performative (Mouritsen, 2006; Diedrich et al., 2013) and interpretative (Easterby-Smith et al., 2012; Sandberg, 2005) stance, because our goal was not to define design thinking or prove its value. Instead, our focus remained on exploring the phenomenon in a particular organizational context, and in building a nuanced understanding of how teams foster design thinking mindsets, and implementing innovation methodology in the case of new product development.

According to Schwandt (2001), qualitative research covers an array of techniques seeking to describe, decode, translate, and come to terms with the meaning, rather than the measurement or frequency of phenomena in the social world. Qualitative research can be interpretive or positive. Following Klein and Myers (1999), the main assumption for interpretive research is that knowledge is gained through social constructions such as language, consciousness, and shared meanings. Additionally, interpretive research acknowledges the intimate relationship between the researcher and the subject. In our case, one of the co-authors participated as a team member in the product development process analyzed in this study. Practice-based research strategy falls under the performative research paradigm, which some authors see as an alternative to qualitative and quantitative paradigms (Haseman, 2006). Practice-based research methods were designed to understand what Donald Schon calls "the situations of practice—the complexity, uncertainty, instability, uniqueness and value conflicts which are increasingly perceived as central to the world of professional practice" (Schön, 1983: 14). Practice-based research is focused on "improvement of practice, and new epistemologies of practice distilled from the insider's understandings of action in context" (Haseman, 2006).

We employed a combination of methods to record, manage, and analyze the data: reflective practice, participant observation, performance ethnography, as well as a series of in-depth qualitative interviews with team members and key stakeholders involved in the project. Ten interviews were selected based on project roles and serve as the basis for a detailed understanding of the person—their views, observations, and vision of the future. By combining the efforts of several interviewees working in a shared corporate environment, a relatively objective view of the situation in which they operate can be obtained. They are therefore well suited for research that requires an in-depth understanding of sensitive phenomena, systems, processes, or personal experiences (Ritchie et al., 2013).

Interviews were conducted with key stakeholders including end users (medical personnel), distributors, product managers, designers, and development engineers.

Each of the interviewees represents a key stakeholder in a particular stage of development following the design thinking model, namely user empathy, problem framing, defining input requirements, ideation, prototyping, and user testing. The interviews focused on the implementation of different phases of design thinking, the role of different stakeholders and their experience, as well as their responses to challenges emerging during product development.

9.4.2 The Company and Design Brief

MESI Medical is a young and fast-growing innovative company headquartered in Ljubljana, Slovenia. It was established in 2010 and has over 50 employees. It is dedicated to the development of non-invasive diagnostic medical devices for use in primary and secondary healthcare. The company's mission is to simplify diagnostics and help discover diseases earlier. The company aims to simplify complex and expensive diagnostic devices designed for early detection and monitoring of various diseases.

They were the first company to launch the "simplest and fastest" MESI ABPI MD ankle-brachial index measuring device that can reliably and quickly diagnose peripheral arterial disease (PAD). Since its launch, MESI ABPI MD has changed the way general practitioners think about checking for the presence of peripheral arterial disease. Prior to MESI ABPI MD, doctors diagnosed the presence of the disease using outdated, slow, and unreliable Doppler probe technology, which required additional staff, time, and training. The basic operating principle of the MESI ABPI MD is simple, so the diagnostic procedure can be performed by virtually anyone with minimal training. The measurement time is only one minute, and after a simple application of blood pressure cuffs, the device independently performs the entire procedure, displaying the result at the end of the test. In comparison, Doppler measurements can take up to 30 minutes and are also more unreliable. This is due to its inability to measure blood pressure in all four limbs simultaneously, which is what is used as the basis for calculating the ankle-brachial index (ABI).

Another important MESI product is the MESI mTABLET diagnostic system, which consists of a medical tablet, wireless diagnostic modules (ECG, spirometer, blood pressure monitor, blood oxygen saturation monitor, etc.), and a patient recording system on a PC. The whole system is based on the idea that the user can operate each of the selected diagnostic modules wirelessly through a single tablet, viewing the results, later, on the tablet or on a PC. The user advantage is that everything is based on a single user interface, which means that the user hardly needs any additional training when adding a new diagnostic module to their routine.

For the purpose of our study, we followed and analyzed the development of a diagnostic module for the measurement of the toe-brachial index (TBI), an additional measurement to the ankle-brachial index. In certain patients at risk of peripheral artery disease, the ankle-brachial index is not entirely reliable because it cannot accurately measure ankle blood pressure due to calcification of the artery walls in the legs. Hardened (calcified) arteries are most common in kidney patients and diabetics, who account for nearly 30% of all patients with peripheral arterial disease (Dhaliwal & Mukherjee, 2007). Since the smaller arteries located in the toes are not prone to mediocalcinosis, medical practice dictates that a toe-brachial index measurement be performed in these patients. The toe-brachial index is measured with three wireless modules—two cuff modules on the arm (which the company has already developed and are identical for the ankle-brachial index) and one that measures pressure in both thumbs with two smaller cuffs and sensors. The toe-brachial index represents the quotient of the systolic pressure of both thumbs and the highest systolic pressure of both arms.

The main challenges in the development of the toe systolic pressure module that the team had to face involved deep understanding of the needs of its users, which in this case are medical professionals who needed to perform the measurement of the toe-brachial index and the reliability and technical performance of the final product design. An additional challenge was the fact that all MESI mTABLET modules are based on the same design language and therefore use almost identical aluminium housings, which are relatively small. This means that development engineers have to find a way to use completely different electronics and hardware components specific to different diagnostic modules in the same footprint.

9.4.3 Motivation for Using Design Thinking for Product Design

The company MESI like its competitors in the EU market was forced to embrace at least some of the principles of design thinking due to the European Union's umbrella document, which lays out the conditions for certification of medical devices on the market. The importance of usability testing in the development of medical devices has been emphasized and enforced since 1993 when the European Council Directive 93/42/EEC concerning medical devices (MDD 93/42/EEC) came into force. In 2017, the European Commission enacted a new document, Regulation (EU) 2017/745 of the European Parliament and of the Council of 5 April 2017 on medical devices, amending Directive 2001/83/ EC, Regulation (EC) No 178/2002, and Regulation (EC) No 1223/2009 and repealing Council Directives 90/385/EEC and 93/42/EEC (MDR 2017/745). The Regulation differs from the Directive in that it is directly applicable in the legislation of the Member States of the European Union, and the legal requirements provided for in the Regulation can be further strengthened by the Member States. MDR 2017/745 therefore imposes even stricter controls on the development and sale of medical devices, which has also influenced the development of a new diagnostic module at MESI. As there was a transition period between 2017 and 2020 when both the old directive and the new regulation were in force, the diagnostic module for measuring the toe-brachial index was the first module developed at MESI in line with the new, stricter regulation.

The new regulation requires more intensive testing of prototypes with users during the development phase, and more thorough product monitoring after its market launch, as specified in the "Post Market Surveillance" document. The latter is also intended to serve as a basis for improvements to an existing product or for the development of new products that are comparable to an existing one. Medical device manufacturers are subject to surveillance by non-autonomous notified bodies and competent authorities. An example of the latter in Slovenia is the Agency for Medicinal Products and Medical Devices of the Republic of Slovenia (JAZMP), and an example of a notified body is the Slovenian Institute of Quality and Metrology (SIQ). Before a product is placed on the market, both bodies must confirm the suitability and safety of the product. This is also a necessary condition for obtaining a CE certificate, which is the entry ticket for selling a medical device on the European market.

MDR 2017/745 requires, among other things, the maintenance of a technical product file containing all documents related to the development of the product. It is worth noting that the content of the technical file and the content of the documentation have a defined indicative format and are not subject to interpretation by the individual manufacturer. The main documents in the technical file include user requirements, risk analysis, feasibility study, clinical study, functional requirements, intended use of the product, and component data sheets. The product manager is usually responsible for the accuracy of the content of the folder in accordance with MDR 2017/745. User testing is usually carried out by the designer and the product manager, who write the user requirements document, from which the functional requirements must necessarily be derived. Starting from the functional requirements that define the final functionality of the product, the product manager must analyze possible risks associated with using the device.

However, the legal requirements were not the sole reason MESI invested in developing internal capabilities for user-centred innovation or why they leveraged design thinking processes in product and UX design. As Steve Eichmann (Nilsson & Sheppard, 2018), who is responsible for design and usability across the Johnson & Johnson Medical Devices Companies, said,

> In the past, some device companies assumed that complying with user-testing regulation and guidelines was enough to make them user-centric. That might have been true up to a point, but laws and regulations don't help you keep pace with design innovation, usability, software development, or technology advances.

Most medical device companies are engaging users in their development beyond legal requirements. For example, at GE Healthcare, empathy was a key requirement to understanding how children experience CT, X-Ray, and MRI scanning procedures. Home visits, family interviews, and sessions with psychologists to understand what paediatric patients go through were carried out to develop a complete understanding. This led to a radically new project framing, which involved more than technical requirements and centred on children's experiences, ultimately leading to an incredibly successful Adventure Series.

At MESI, it was clear that in addition to the ease of use, clinical efficacy, and safety, they need to focus also on better outcomes, lower costs, and patient satisfaction. To meet these needs, designers and usability experts on the MESI team needed to reduce device complexity and variability and improve their reliability, patient outcomes, and overall performance. That was achieved by nurturing design thinking mindsets, following design thinking processes, and engaging key stakeholders in the development process.

9.4.4 Design Thinking Step-by-Step: Chronological Overview of a New Medical Device Development

The development of the MESI mTABLET TBI module was based on the validated requirements for the company to start developing a diagnostic module for the measurement of the toe-brachial index—the next step after the ankle-brachial index in the diagnosis of peripheral arterial disease in high-risk patients. As such, the module fits well into the company's existing product portfolio, which includes two ankle-brachial index monitors—one as a standalone device and the other as a diagnostic module within the MESI mTABLET system.

Product development was initiated by the product manager with a feasibility study, as envisaged in the technical dossier scheme derived from MDR 2017/745 and the harmonized standard ISO 13485. The feasibility study started with an estimate of the timeframe required for the project, which was the base document that allowed the product manager to prepare a preliminary product development plan. The development plan covers all phases of development and serves as the basis for the project manager to assign tasks to the development staff. Yet, it also acts as a commitment to the executive team and the sales and marketing departments as to when the product will be ready for sale. They can then officially inform other stakeholders (distributors, retailers, customers, etc.) when the product will be available.

Feasibility study is to some extent a "living" file that is updated throughout the process. However, it is important that the documentation is as detailed as possible from the beginning stages of the development. Once the feasibility study has been validated, the product manager holds a meeting with senior management to present the findings. At this meeting, management gives its final approval to product development, as the feasibility study is a kind of preliminary phase where only the product manager contributes his part without input from other development staff. During the feasibility study, the product manager conducted interviews with specific stakeholders (in this particular case, healthcare professionals) to better understand the market, competing devices, disease states, and measurements. In our case, the feasibility study confirmed the initial assumptions, and management confirmed that the development department should proceed with the development of the module. The feasibility study was also validated shortly after its completion by a notified body with whom MESI was working as part of the announced external audit for compliance with ISO 13485. Once the feasibility study was completed, the product manager, in collaboration with the project manager and the head of the product office, prepared a schedule document for the implementation of each phase of the project. This document differs from the project schedule estimate in that it specifies the exact time frame for the smaller phases. So, it is roughly a more detailed document serving the same purpose.

Next came the writing of the user requirements document, which defined, point by point, all the functionalities that the product must contain or, in other words, the functionalities that the end user expects from the product. The user requirements document defined the general physical characteristics of the product, such as dimensions, materials, and colours, as well as its functional characteristics, such as the way the battery is charged, the length of the hoses and cables, the volume of the attachment mechanism, the colour of the warning LEDs, and the volume of the alarms. The user requirements also encompass the software part of the device and define the functions available to the user on the MESI mTABLET tablet: the measurement scales, the units used, the size of the letters, the order in which the information is displayed.

Writing user requirements is also the phase where the product manager and user experience designer begin to empathize with the user by conducting user interviews. Product managers and designers from MESI whom we interviewed for this study pointed out the parallels with the theoretical model of design thinking in this phase, agreeing that it is an iterative process that typically combines the phases of user empathy and problem definition. They do this in meetings with existing and potential users that can last up to several hours, sometimes in the form of brainstorming sessions, where they try to determine pain points and explore what could help them do their jobs more efficiently and effectively. If they are unsure about their assumptions, they check them with the help of additional medical professionals they already work with. Based on these findings, the product manager can finalize the user requirements document. The standard requires that these documents be updated on an ongoing basis.

The user requirements document served as the basis for writing the functional requirements, which were written by the mechanical developer, the electronics developer, the hardware developer, and the software developer, each for their own specific area. These defined specific product characteristics such as material type and surface, material surface treatment, degree of waterproofing, component set, software code language, and communication protocol.

The validated functional requirements served as an important milestone for the first prototypes. MESI tests prototypes on multiple levels—mechanical components are typically tested with 3D printed parts, and electronic performance is tested with soldered components (final PCBs are not yet manufactured). They tested the design on a tablet in multiple stages—the first designs were tested with users utilizing only printed design templates on paper. In the second phase, they used a web-based tool that allows them to easily programme certain design elements to become clickable and illustrate the user process. Only in the final prototyping phase they actually programmed the design, tested it, and adjusted it as needed. All interviewees agree that one of the main problems with prototyping is that no approximation is identical to the final piece and therefore cannot 100% reflect how the final design will behave. Therefore, the only real indication of the quality of the assumptions is still the final product. Nevertheless, our interviewees agreed that prototyping is a valuable process that allows them to develop higher quality products and gives them confidence in the accuracy of their assumptions during the development process. Miscalculations can be very costly to a company, as retrofitting injection molds, for example, can cost tens of thousands of US dollars. Three-dimensional printed parts allow designers to quickly and inexpensively produce prototypes that are similar in function and form to the final version.

Such a prototyped device can serve as the basis for user experience testing by a product manager and a designer. At MESI, prototype devices are typically assembled from 3D-printed and finished components that can be taken from existing devices. The electronics designer prepares prototype circuitry that is implemented in the prototype, and the hardware designer ensures that the device is able to function virtually identically to the final product in its best iteration. With varying degrees of sophistication, the product manager and designer visit the end users and provide guidance to the developers on changes and improvements based on their feedback. In the specific example of developing a diagnostic module to measure finger pressure, it has proven virtually impossible to predict the testing schedule for prototype devices. During the development of this device, MESI produced at least 10 prototypes over a 12-month period before they could confirm the suitability of the design. During this time, they tested and modified nearly every component of the original concept and worked with potential users to bring it to a level that provided a significantly higher level of satisfaction than the original design.

In the testing phase of the physical prototypes, it is also worth mentioning the importance of user feedback, which is not only communicated to the developers, but is also taken into account in the device risk analysis. In the course of this analysis, the product manager must anticipate, evaluate, and address all the risks that the use of the device may bring. The risk analysis document was derived from the preliminary risk analysis document, which is part of the feasibility study. It is not until the user testing phase that the product manager can actually assess the likelihood and severity of each risk. If the latter is too high, the design of the device must be adjusted, regardless of end-user opinions. If the risk cannot be reduced, it must be addressed in other ways—warnings, instructions, additional user training, etc.

Once the product is defined in terms of design and function, it is also ready for the laboratory and safety tests carried out by the notified body. The latter is responsible for verifying that the product is safe and suitable for use in critical industries such as medicine. The fact that legislation requires usability tests is proof that usability is not a design "myth" but a real feature of the device that in principle costs the company nothing, apart from the time it invests in finding the best possible functionality for the device. This is defined by the IEC 62366–1:2015 Medical devices—Part 1: Application of usability engineering to medical devices.

Once a device's suitability was finally confirmed by a notified body, the engineering department prepared detailed instructions for assembly and passed the documentation to the manufacturing departments. After market launch, the product is monitored on the market in compliance with the law as part of the follow-up inspection, which also serves as a source of information for improvements to the existing product, and the development of further products. In this sense, the product never leaves the testing phase, and the product manager has a working database with ideas for potential improvements based on constant user feedback.

9.4.5 Key Benefits and Challenges When Using Design Thinking as a UX and Product Design Tool

Based on the interviews with MESI employees, testing prototype devices with end users has been one of the cornerstones of new product development. Not only was testing necessary to meet regulatory requirements, but it has also proven to be an effective process in new product development. As the company started to introduce design thinking into their product development process few years ago, it became apparent that the process itself and the execution of each phase needed to be more systematized and documented. According to our interviews, a more systematic approach was specifically important during the user research and user testing phase. Company's user experience (UX) designer, whose main task is to make the user interface of the MESI mTABLET device as simple and pleasant as possible, while enabling the quickest possible diagnosis, started introducing more extensive and systematic user testing. She played an important role during this project by developing a better organized internal system that follows the design thinking process. According to this newly implemented system, the company only allows adoption of solutions that have been extensively tested with users. In exceptional cases when testing with end users is not possible, the internal system requires a new feature to at least be tested internally with a larger group of people who can provide an independent response.

This new user-centric development and innovation process that MESI adopted was greatly appreciated by the end users, which in this project were medical professionals. They were pleased that someone was listening to them and showed genuine care for their experience while using the product. As one doctor described it, "this is how one can develop devices that do not just serve the manufacturer's profit but actually improve our work and patients' lives". Another doctor aptly highlighted the importance of innovation in medical devices, "doctors are just highly skilled workers working with the tools at their disposal, and it is the medical device manufacturers who have the power to make real progress".

An interesting spillover effect of applying design thinking is also evident in the positive attitudes of retailers. They have been increasingly involved in product development, at least in terms of directly relevant content such as the size and shape of transport packaging, and the layout of EAN labels. This way of working has given them the signal that they are being heard and that their experience matters. The benefits of the newly adopted innovation approach were also recognized by the company's management and sales staff, who have been receiving increasingly positive feedback from their partners. The positive reactions from the end users and other stakeholders gave company an increased confidence that user-centred approach to innovation is paramount in developing new solutions.

It also gave them courage and motivation to continue to deepen their design thinking capabilities and make their innovation process even more systematized and robust. They hired a second user experience designer who joined the company at a time when MESI was conducting intensive testing to develop new applications for the MESI mTABLET system. Experience design team developed a user experience test protocol that now serves as the basis for all further tests, which are carried out by product managers with the assistance of medical professionals when developing new devices.

Product managers are also responsible for testing the physical parts of the device, not just the user interface on the tablet. For example, the company works with mechanical designers to develop and test various design solutions that can be tested against 3D-printed models. The company does not have an in-house industrial designer and works with an outside contractor for more complex projects. For smaller designs, a mechanical designer and product manager bring the idea to life. Three-dimensional printers allow for rapid prototyping, which can be used to test the functionality of an imagined part within a few days. MESI does not yet have a 3D printer, although it admits that it would pay for itself given the number of pieces it has ordered so far. This shows that in this area, too, employees test various ideas, which are finally approved only after they have been validated by users and meet the functional requirements. As evident from Figure 9.4, the outcome of using design thinking is a device that showcases an array of advantages, especially when compared to other devices available on the market.

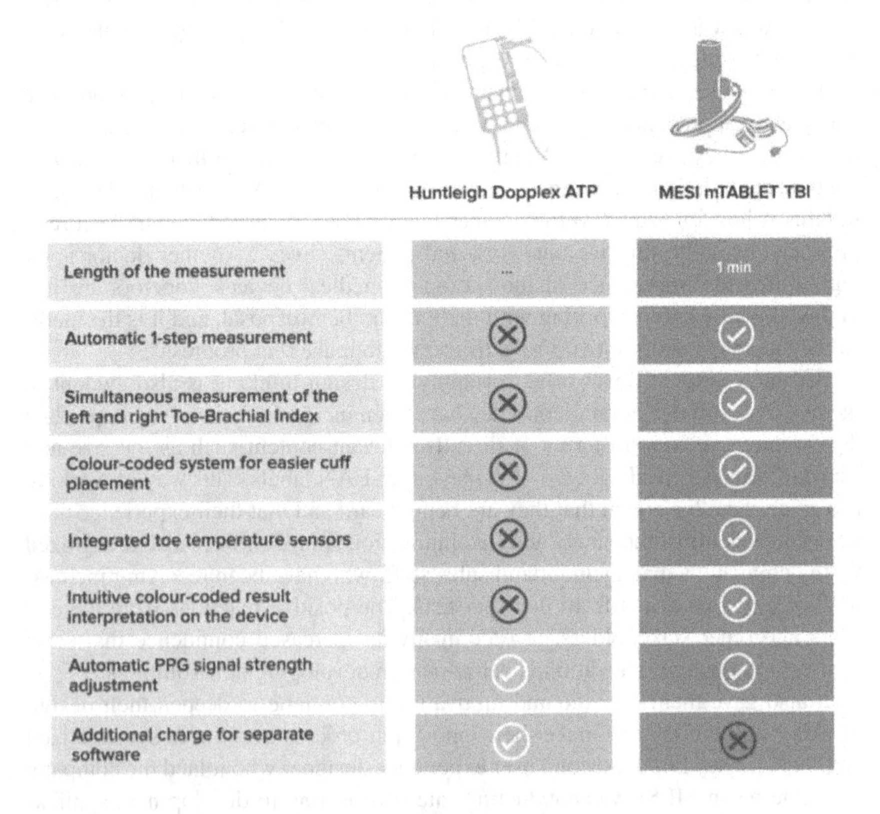

	Huntleigh Dopplex ATP	MESI mTABLET TBI
Length of the measurement	...	1 min
Automatic 1-step measurement	⊗	✓
Simultaneous measurement of the left and right Toe-Brachial Index	⊗	✓
Colour-coded system for easier cuff placement	⊗	✓
Integrated toe temperature sensors	⊗	✓
Intuitive colour-coded result interpretation on the device	⊗	✓
Automatic PPG signal strength adjustment	✓	✓
Additional charge for separate software	✓	⊗

FIGURE 9.4 Benefits of MESI mTABLET TBI (courtesy: MESI d.o.o.).

The mechanical engineer we interviewed mentioned that sometimes user needs can lead to ideas that are not feasible, and he highlighted the importance of collaboration among mechanical engineers, product managers, and end users to find solutions that lie in the intersection of user needs (desirability), feasibility, and viability—and will prove profitable for the company at the end of the project.

9.5 DISCUSSION

According to the data we gathered during the product development process and with in-depth interviews, we can conclude that introducing design thinking into the product development process has been successful and will help the company remain globally competitive and maintain its position as an innovation leader within their market niche. Close contact with users and engaging diverse stakeholders in the development have increased motivation and enthusiasm of designers and developers as they provided them with ongoing feedback and ability to move to the next stage of development with conviction that they will develop a desirable and effective product.

When asked about the advantages and disadvantages of using design thinking, a product manager at MESI responded:

> The main advantage of using design thinking is combining diverse perspectives and exploring diverse ideas because every team member brings a diverse and unique perspective. But this diversity can also be a disadvantage because it is sometimes difficult to converge and decide on one idea out of a multitude of different ideas. Each individual has their own subjective view and it is difficult to know which idea is the best.

Based on our observation, we can conclude that despite providing a share and clearly defined process, design thinking needs to be combined with a systematic project management. To effectively manage a project and to adhere to the project development timeline, it is important to provide timeframes within which each phase or a cycle needs to happen. Apart from a clearly defined timeframe, it is also critical to clearly define roles in each stage of the process and effectively manage the hand-offs between stages and team members. This allows the development team to remain accountable to the sales and marketing departments and to bring the product to market on time.

Another important aspect that was raised during the interviews was the importance of introducing design thinking capabilities and mindsets to the rest of the company and not keep it narrowly used only by the product development team. The user experience designer emphasized the importance of "applying design thinking holistically within a company and increase the awareness of the benefits that high-quality and consistent user testing can bring". A mechanics developer at MESI illustrated that by saying that "the team often felt rushed due to the project's timeframe, which was dictated by the sales department". This suggests that while the company as a whole acknowledges the benefits of the user-centred innovation approach, not everyone in the organization is aware that this type of product development requires longer time in certain phases due to qualitative user testing, which can be time and energy consuming. A well-known saying that summarizes this is "going slow to go fast". It means that with this approach to innovation, you sometimes spend extensive time in certain phases which at the first sign might be seen as slowing down the project but in reality it helps the team to accelerate the overall progress.

Another challenging aspect of using design thinking that was often hard to explain to the internal stakeholders was the iterative nature of the process. At MESI, they follow all the phases of the theoretically proposed design process, but the biggest frustration and misunderstanding among colleagues were caused when testing and user feedback indicated that the team needs to return back to the earlier phases of the process. This iterative approach is one of the main aspects of design thinking, but, based on interviews, it often causes conflicts with a more linear project management approach and is still something the company is trying to resolve. We believe that the reasons for the misalignment of phases lie in the fact that the development of mTABLET modules is a complex product that requires the fulfilment of development requirements at several levels. Testing with users must be performed both at the level of the diagnostic device itself and at the level of the user interface on the tablet. At the same time, the user interface must be unified with the applications of the other diagnostic modules, which in some cases leads to contradictions, as the diagnostic devices and their functions are often not equivalent. In addition, the development of a medical device also requires the validation of clinical values with experts whose opinions often differ.

Nevertheless, the interviewees believe that the management of the development stages is improving with the recruitment of a project manager responsible for tracking and coordinating all the stakeholders involved in the development. In our opinion, the situation could be further improved by educating other employees more about what design thinking is, and, more importantly, what its benefits are. This is an opportunity for breaking the silos that have long been established in medical device companies and to foster true "radical collaboration". If all employees understand how the process works, how different phases follow each other and sometimes overlap, and why certain phases take a certain amount of time, the work can be better planned ahead of time and can be less stressful for all those involved. The process of design thinking consists of individual phases that come together to form a process cycle, and during a new product development the team proceeds through multiple cycles. The process cycle is described in the literature not as a sequence of steps but rather as a process comprising varying phases. Therefore, it is important to view design thinking as a system of overlapping phases. The justification for this is that they constantly flow into each other. Ideas can always alternate between the prototype phase and the inspiration or ideation phase. Subsequently, it is not surprising that design thinking can appear chaotic to an outside observer or a new member of the design team (Brown & Wyatt, 2010), and it is crucial to have a shared understanding within the company of how and why the process evolves.

9.6 CONCLUSION

In situations when doctors have to make important decisions at a moment's notice, a well-designed medical device can make the difference between life and death, a fact that influences the motivation of those involved in the development of such important products. However, the motivation and knowledge of those designing medical equipment are not enough. One key element that is needed to design a medical device that is not just effective, but also desirable and offers a superior user experience is empathy for the people using it. One of the innovation methodologies that is deeply rooted in empathy is called design thinking, and it allows the design team to gain a deep understanding of explicit as well as underlying user needs. In design thinking, users are involved in the innovation process from the very beginning through ethnographic user research and then again at the end of each innovation cycle during user testing.

While extensive user testing is required by law before each medical device becomes certified, more and more medical device companies are realizing the value of engaging users already at the forefront, when they are exploring user needs, identifying pain points in the current user experience, and defining project requirements. Design thinking offers companies (1) a well-defined process, (2) common language, and (3) appropriate tools to use in each of the development phases. These three aspects are especially important when working on a complex project in cross-functional teams and when collaborating and co-creating with stakeholders who are not directly involved in the actual product development. Internal stakeholders include other departments in the company, such as sales, marketing, and production, while external stakeholders include distributors, resellers, users, and component suppliers.

Our analysis shows how design thinking can be successfully applied as a product design methodology and illuminates why a user-centric approach to innovation is extensively used in the medical device manufacturing sector.

REFERENCES

Brenner, W., Uebernickel, F., & Abrell, T. (2016). Design thinking as mindset, process, and toolbox. In *Design Thinking for Innovation* (pp. 3–21). Cham: Springer.

Brown, T. (2008). Design thinking. *Harvard Business Review*, 86(6), 84–92.

Brown, T., & Wyatt, J. (2010). Design thinking for social innovation. *Stanford Social Innovation Review*, 8(1), 30–35.

Buchenau, M., & Fulton Suri, J. (2000). Experience prototyping. In *Proceedings of the Conference on Designing Interactive Systems* (pp. 424–433). New York: Association for Computing Machinery.

Chasanidou, D., Gasparini, A. A., & Lee, E. (2014). *Design Thinking Methods and Tools for Innovation in Multidisciplinary Teams*. Cham: Springer.

Dam, R., & Siang, T. Y. (2018). *What is design thinking and why is it so popular*. Aarhus: Interaction Design Foundation. https://athena.ecs.csus.edu/~buckley/CSc170_F2018_files/What%20is%20Design%20Thinking%20and%20Why%20Is%20It%20So%20Popular.pdf

De Bono, E. (2017). *Six Thinking Hats: The Multi-Million Bestselling Guide to Running Better Meetings and Making Faster Decisions*. London: Penguin.

Derenda Mujezinović, A. (2017). *Načrtovanje digitalnih prodajnih procesov medicinskih naprav in aplikacij (masters thesis)*. Ljubljana: Naravoslovnotehniška fakulteta.
Design Council (2007). The value of design factfinder report. Obtained on 12 December 2019 from www.designcouncil.org.uk/sites/default/files/asset/document/TheValueOf DesignFactfinder_Design_Council.pdf
Dhaliwal, G., & Mukherjee, D. (2007). Peripheral arterial disease: Epidemiology, natural history, diagnosis and treatment. *International Journal of Angiology*, 16(2), 36–36.
Diedrich, A., Eriksson-Zetterquist, U., Ewertsson, L., Hagberg, J., Hallin, A., Lavén, F., ... Walter, L. (2013). Exploring the performativity turn in management studies. Gothenburg: Gothenburg Research Institute.
Dorst, K. (2011). The core of 'design thinking' and its application. *Design Studies*, 32(6), 521–532.
Dow, S. P., Glassco, A., Kass, J., Schwarz, M., Schwartz, D. L., & Klemmer, S. R. (2010). Parallel prototyping leads to better design results, more divergence, and increased self-efficacy. *ACM Transactions on Computer-Human Interaction (TOCHI)*, 17(4), 1–24.
Dunne, D. (2018). Implementing design thinking in organizations: An exploratory study. *Journal of Organization Design*, 7(16), 1–16.
Easterby-Smith, M., Thorpe, R., & Jackson, P. R. (2012). *Management Research*. Thousand Oaks: Sage.
Fry, T. (2009). *Design Futtering Sustainability, Ethics and New Practice*. Oxford: Berg
Gregersen, H. (2018). Better brainstorming. *Harvard Business Review*, 96(2), 64–71.
Hammersley, M. (2016). *Reading Ethnographic Research*. Abingdon-on-Thames: Routledge.
Haseman, B. (2006). A manifesto for performative research. *Media International Australia*, 118(1), 98–106.
Houde, S., & Hill, C. (1997). What do prototypes prototype. In: M. Helander, T. K. Landauer, & P. Prabhu (eds.), *Handbook of Human-Computer Interaction* (pp. 367–381). North Holland: Elsevier.
Interaction Design (2020). 5 Stages in the design thinking. Obtained on 2 February from www.interaction-design.org/literature/article/5-stages-in-the-design-thinking-process
Klein, H. K., & Myers, M. D. (1999). A set of principles for conducting and evaluating interpretive field studies in information systems. *MIS Quarterly*, 67–93.
Kolko, J. (2014). For any product to be successful, empathy is key. *Harvard Business Review*, 18 December 2014.
Leonard, D., & Rayport, J. F. (1997). Spark innovation through empathic design. *Harvard Business Review*, 75, 102–115.
Liakhavets, A., & Petukh, A. (2019). *Barriers to and Facilitators of Design Thinking Implementation in Companies (masters thesis)*. Ljubljana: University of Ljubljana, School of Economics and Business.
Liedtka, J., & Ogilvie, T. (2011). *Designing for Growth: A Design Thinking Tool Kit for Managers*. New York: Columbia University Press.
Lojacono, G., & Zaccai, G. (2004). The evolution of the design-inspired enterprise. *MIT Sloan Management Review*, 45(3), 75–79.
Lucas, B. J., & Nordgren, L. F. (2020). The creative cliff illusion. *Proceedings of the National Academy of Sciences*, 117(33), 19830–19836.
Martin, R. L. (2014). The unexpected benefits of rapid prototyping. *Harvard Business Review*.
Martin, R. L., & Martin, R. L. (2009). *The Design of Business: Why Design Thinking is the Next Competitive Advantage*. Boston: Harvard Business Press.

Mouritsen, J. (2006). Problematising intellectual capital research: Ostensive versus performative IC. *Accounting, Auditing & Accountability Journal*, 19(6), 820–841.

Mural (2018). Evaluating the impact of design thinking in action: Webinar recap. Obtained on 20 January 2020 from https://blog.mural.co/designthinking-roi

Nilsson, T., & Sheppard, B. (2018). *The Changing Face of Medical Device Design.* London: McKinsey Insights.

Pinder, M. (2022). 16 Cognitive biases that can kill your decision making. *Board of Innovation.* Obtained on 4 February 2022 from www.boardofinnovation.com/blog/16-cognitive-biases-that-kill-innovative-thinking/.

Ritchie, J., Lewis, J., Nicholls, C. M., & Ormston, R. (Eds.). (2013). *Qualitative Research Practice: A Guide for Social Science Students and Researchers.* London: Sage.

Rosenberg, N. O. Sr., Chauvet, M. C., & Kleinman, J. S. (2015). *Design Thinking: New Product Development Essentials from the PDMA.* Hoboken: Wiley.

Sandberg, J. (2005). How do we justify knowledge produced within interpretive approaches? *Organizational Research Methods*, 8(1), 41–68.

Schmiedgen, J., Rhinow, H., Köppen, E., & Meinel, C. (2015). *Parts Without a Whole: The Current State of Design Thinking Practice in Organizations.* Potsdam: Hasso-Plattner-Institut für Softwaresystemtechnik an der Universität Potsdam.

Schön, D. (1983). *The Reflective Practitioner – How Professionals Think in Action.* New York: Basic Books.

Schwandt, T. A. (2001). *Dictionary of Qualitative Inquiry.* Thousand Oaks: Sage.

Seidel, P. V., & Fixson, K. S. (2013). Adopting design thinking in novice multidisciplinary teams: The application and limits of design methods and reflexive practices. *The Journal of Product Innovation Management*, 30(S1), 19–33.

Simon, H. A. (1988). The science of design: Creating the artificial. *Design Issues*, 4(1), 67–82.

Sutton, R. I. (2007). The power of the prototyping mind-set. *Harvard Business Review.* Obtained on 2 February 2022 from https://hbr.org/2007/05/the-power-of-the-prototyping-m-1.

Sutton, R. I., & Hargadon, A. (1996). Brainstorming groups in context: Effectiveness in a product design firm. *Administrative Science Quarterly*, 685–718.

Zupan, B., & Svetina Nabergoj, A. (2014). Razvoj podjetniških kompentec s pomočjo dizajnerskega pristopa. *Economic and Business Review*, 16, 49–74.

10 Automatic Coin Counter and Sorter—Design Thinking Approach for Prototyping

Senthil Vinod S., Sibi Kumar M., Siva Subramaniyen M., Sudharshana Venkatesh C., Tharun K. M. and D. Kavitha
Department of EEE, Thiagarajar College of Engineering, Madurai, India

DOI: 10.1201/9781003189923-13

10.1 INTRODUCTION

India contains a large number of coins in circulation in the financial and other fields; this coin inventory and classification are done through manual means, and this mode of operation and low work efficiency give the operator a great trouble, as the labor intensity is great and not accurate. The systems of the machine that are available in the market are not good enough because sometimes they make mistake when detecting the value of coin and cause the error calculation in the result. This is mentioned in the context of Malaysia [1]. Counting and interpreting currency coins with accuracy and high speed are a challenging problem for banks and stores and even for consumers [2]. The design of the microcontroller as the control center, combined with the sensor technology, and the corresponding mechanical structure to complete the separation of coins and finishing the packaging work and real-time monitoring and display of the type and number of coins function are explained well in the literatures [3–4]. The methods used in coin recognition, such as coin detection and extraction, feature extraction, classification, and verification, are also dealt in detail in literatures [5]. Even system hardware and software design suffers problems of reliability [6].

Table 10.1 defines the members involved in projects and the roles played by them.

TABLE 10.1
Name of the Team and Roles and Responsibilities of the Team

S. No.	Name	Roles and Responsibilities
1	Member 1	**Leader and Organizer**—one who allocates work to the team and organizes meetings
2	Member 2	**Task Executer and Web master**—one who looks after time management and collects information from web
3	Member 3	**Project Manager**—one who manages the financial part
4	Member 4	**Designer**—one who designs prototypes/solutions and communicates with the stakeholders
5	Member 5	**Designer**—one who designs prototypes/solutions and communicates with the stakeholders

This project work explained in this paper will come under the ninth sustainable development goals as it can have huge market value because of its demand, time saving characteristic and cost-effectiveness as shown in Figure 10.1 [7]. Till now there are no low-cost products to address the problem [8]. The proposed product can count and sort coins at a low cost. The resulting product will end the current problems and will be an invaluable asset for those who work with coins in their daily life and avoid the complexity involved in human sorting as illustrated in Figure 10.2 [9]. Societal need of the solution to the problem is understood from the survey taken using Google form and is given here:

1. This project will be helpful for the shopkeepers who deal with coins in each and every sale as our product will automatically sort and count the coins. This will save a lot of time for them and they can easily give reminders to their customers without any hassle.
2. The proposed project will be helpful to the temple authorities for counting the coins.

Motivation for the planning and execution of the project is obtained from one of the stakeholders, who is a financial businessman mostly dealing with coins and currency and finding difficulty in counting coins.

FIGURE 10.1 Sustainable development goal addressed (source: https://sdgs.un.org in 2021).

FIGURE 10.2 Human sorting (source: https://patents.google.com in 2021).

10.2 PROBLEM IDENTIFICATION AND EVALUATION

Effects of counting coins manually may lead to the problems such as increase in labor work, time-consuming, less accuracy, which affect the daily work. This represents the need of the community for: "An automatic coin counter and sorter to make their work easier".

Currently, there are products available in the market to do this job, but their costs are very much higher and cannot be afforded by both the middle- and low-class people who count the coins and sort them manually. They have to invest with their time and man power to do this job daily.

Suitable stakeholders are identified to list out the requirements. The list includes a businessman, temple authority, finance person and bakery shop keeper.

The primary goal of the stakeholders is to create a product at a low cost so that everyone can count and sort the coins automatically.

Tools used to identify stakeholder requirements are face-to-face interviews, brainstorming sessions with group members, and the Google survey form. The consolidation results in the following requirement identification.

Functional: It should be capable of both counting and sorting coins.

Performance: It should give a fast and effective solution to count and sort coins

Physical: It should be portable.

Regulatory: It should be able to be used according to the requirements.

Economical: It should be affordable (specifically around Rs. 1,500).

Environmental: It should be eco-friendly.

The voice of customer is obtained using Google forms. The responses are collected from people who are primarily involved in businesses handling cashes frequently. Simple surveys are conducted where questions such as "Do you feel comfortable using coins in day to day life?", "Which is difficult to count? Coins or Notes", "Do you ever frustrated while counting coins?", "Do you need Coin sorting and counting machine at affordable cost?", etc., are asked.

Figure 10.3 illustrates the ease in using coins among the people nowadays [9].

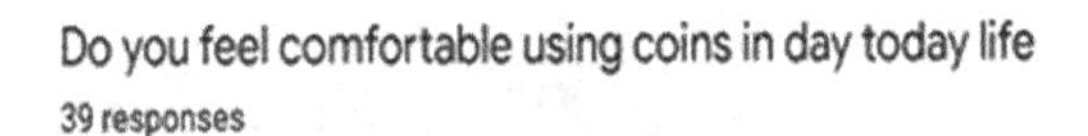

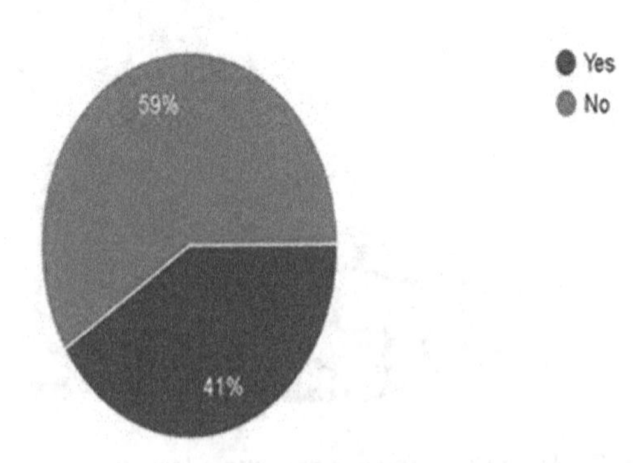

FIGURE 10.3 Handling coins (source: https://patents.google.com in 2021).

From Figure 10.4, it is evident that majority of the people have the opinion that coins are more difficult to count than currencies [9].

Figure 10.5 states clearly that almost 75% people got frustrated while counting coins, and most people require a coin sorting and counting machine at an affordable cost, as is given by Figure 10.6 [9]. It is quite interesting to find that people get irritated of coin sorting and counting process, but they are ready to use coins after they have been counted and sorted.

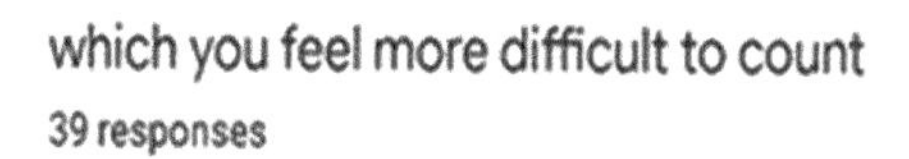

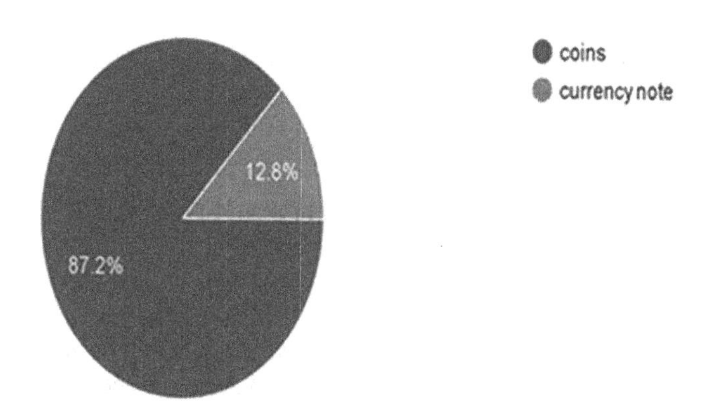

FIGURE 10.4 Difficulty in using coins over currency (source: https://patents.google.com in 2021).

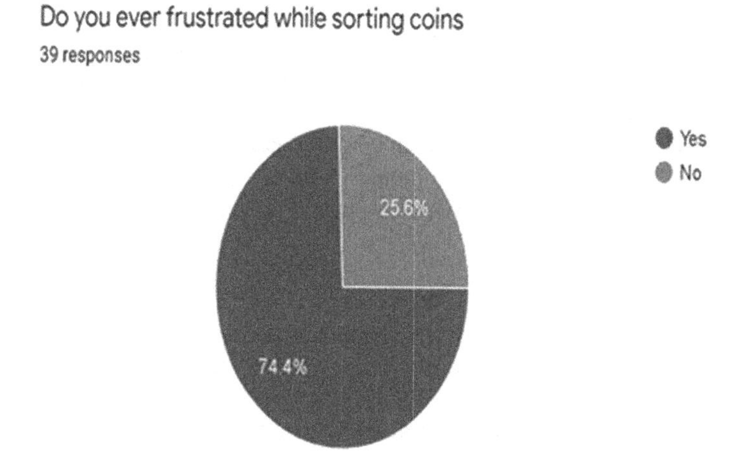

FIGURE 10.5 Frustration while sorting coins (source: https://patents.google.com in 2021).

would you feel comfortable to use coins if there is a machine to do your sorting and counting at an affordable cost

39 responses

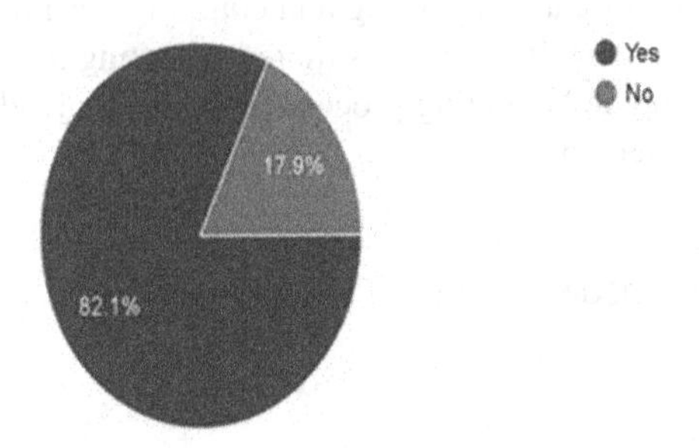

FIGURE 10.6 Need for a counting machine at affordable cost (source: https://patents.google.com in 2021).

10.3 SPECIFICATION DEVELOPMENT

10.3.1 Considerations

While developing the required specifications for the product, it is essential to keep in mind that the product should be real and able to be commercialized. Expectations of stakeholders are collected, noted down, and converted to specifications.

When a suitable solution is implemented, dealing with coins will become much easier. It will decrease the time and man power to be invested in counting and sorting of coins. Our project will satisfy the stakeholder requirements at a reasonable cost.

10.3.1.1 Reality

India is a country which has a large number of middle-class people whose livelihood depend on small shops, business, travel, and so on, and it is a multi-religious country which has a large number of temples. They are always dealing with situations which involve coins rather than currency notes. Exchanging of coins and finding the right coin take some amount of time which will frustrate both the customer and seller. Counting of coins especially in temples requires both a lot of time and man power.

10.3.1.2 Consequences

If this problem is not fixed, both time and labor to be invested in it will keep on increasing. Everyone needs to wait in shops to get their change. Conductors will continue to face the difficulties. Time is a very precious thing which cannot be wasted in waiting and searching for the right penny.

TABLE 10.2
Stakeholders' Expectations

Criteria	Stakeholder 1	Stakeholder 2	Stakeholder 3	Stakeholder 4
Source	It should work in electricity and should have power backups to work in necessary situation.	It should be of rechargeable type.	It should be in electricity	It should be rechargeable
Cost	Should be affordable	Should be affordable	Should be affordable	Should be Low cost
Size	Small and should be portable	Small and should be portable	Large so that more quantity can be counted	Does not matter only efficiency matters
Working	It should be able to separate and count the coins and also packs them separately.	It should separate and count the coins	It should count the coins and should be waterproof	It should separate the coins

10.3.1.3 Proposal

Though science and technology have improved a lot, still it seems to be impossible to serve the last person.

There are many products available in the market, but they are not affordable for a common man. This proposed product will target these needy people. It will be beneficial for the community if the proposed product is made available in every small shops and temples so that everyone can know about this and use it. Table 10.2 specifies the various stakeholder's expectations in the proposed work.

10.3.2 Available Market Products

Several models are available in market. Figure 10.7 shows one of them and is a coin sorting and counting machine from a Paras brand. It can sort and count all types of Indian currency coins [10]. The model (link: www.amazon.in/PARAS-950-Speed-Counting-Sorting-Machine/dp/B07MY43XN3/ref=sr_1_28?keywords=coin+sorting+machine+PARAS-950&qid=1638377890&sr=8-28) is a high-end one that exists in the market and is marketed via Amazon. The basic model starts from INR 25,000. The public review of the same is taken from Amazon and provided in Figure 10.8 [10].

FIGURE 10.7 Existing product, the coin sorting and counting machine (source: www.amazon.in 2021).

Dan N.

Jammed way too easily

Reviewed in the United States on 23 September 2020

Verified Purchase

Colour: Black

Bought this to help process weekly tips at my store. It worked twice before jamming up. Had to buy a different coin sorter, it just wasn't worth the time wasted trying to fix or unjam this every five minutes.

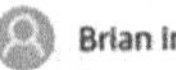

Brian in Langley BC

Verified Purchase

Reviewed in Canada on 8 August 2020

Pretty Darned Quick, and Easy Too!

This little machine sorts coins very quickly, however... You should really use the coin-tubes recommended for this machine, as the inexpensive bulk coin-tubes are not at all meant to be used on this device, and you WILL be recounting every single tube of coins for accuracy.... See more

FIGURE 10.8 Review on Paras brand machine (source: www.amazon.in 2021).

10.3.3 Development of Customer Specification

Customer specifications with respect to goal, scope and constraints are identified.

10.3.3.1 Goal

- To reduce the effort and time while counting coins.
- To create a product that can reach all the needy people.
- To reduce the mistakes while counting coins.
- To make a product in a very affordable cost.

TABLE 10.3
Evaluation Criteria for Identified Parameters

Sr. No.	Parameter	Inference and Evaluation Criteria
1	Demand	There is a growing demand for a product to count and sort coins as currently there are only costly products to do this job. Market survey regarding the demand is taken.
2	Efficiency	Currently, coin counting and sorting are done manually; our product will do it automatically, which will reduce time and human error and for which time is the evaluation criterion.
3	Impact	This will have a huge impact on coin transactions because our product will make the transactions automatic and so everything will be easy. The number of impediments while using are identified and are tested.
4	Durability	Our project is planned in such a way so that the machine can have a long life on the basis of the way it is designed and materials used. Standard durability tests will be done.

- To create a product that is very much handy and portable.
- To denominate the use of currency.

10.3.3.2 Scope

- No need for human intervention
- Less time consumption
- Low cost automation
- High operating accuracy
- Digital display providing high accuracy in results

10.3.3.3 Constraints

- It has a one-time huge investment.
- If delicate sensors are used, they should be always kept in check for getting accurate results.
- Sensor failure may lead to faulty computation.

10.3.4 Development of Evaluation Criteria

Table 10.3 provides the various criteria to be considered and the evaluation criteria to be used.

10.4 CONCEPTUAL DESIGN

10.4.1 Brainstorming of Possible Solutions

Different possible solutions are derived using brainstorming sessions. Few of them are listed here.

1. Counting can be done through a weight sensor (link: www.sparkfun.com/products/10245). The weight sensor is shown in Figure 10.9 [11].
2. Counting can be done through the IR sensor [12] (link: www.indiamart.com/proddetail/ir-sensor-21495675055.html). IR sensor is shown in Figure 10.10 [11].
3. Counting can be done through an image sensor (link: www.novuslight.com/organic-cmos-image-sensor-market_N9398.html). The image sensor is shown in Figure 10.11 [13].
4. Sorting coins is done on the basis of diameter of the coins. This is the only method discussed to sort coins as it meets all our needs.

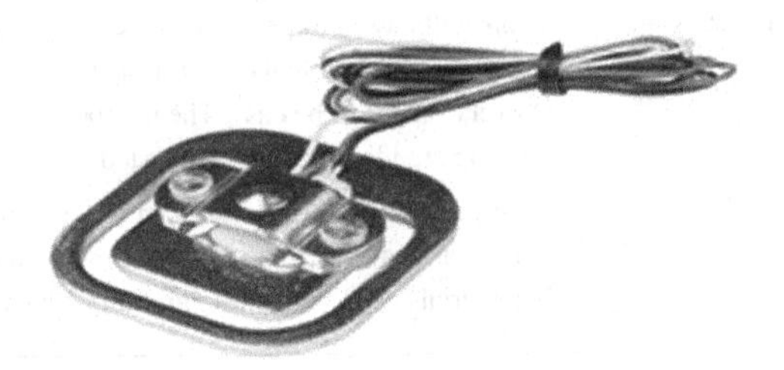

FIGURE 10.9 Weight sensor (source: www.sparkfun.com/products/ in 2021).

FIGURE 10.10 IR sensor (source: www.sparkfun.com/products/ in 2021).

FIGURE 10.11 Image sensor (source: www.novuslight.com in 2021).

TABLE 10.4
Average Score of Potential Solutions

Solution	Cost	Portability	Speed	Accuracy	Total
Weight sensor	9	8	7	7	31
IR sensor	9	9	9	8	35
Image processing	7	8	8	9	32

10.4.2 Evaluation of Potential Solutions

Given in Table 10.4 are the ratings out of 10 for each category. The evaluation is done by each and every member of the group, and the average is rounded to the nearest integer as is provided in Table 10.4.

10.4.3 Selection of Best Design

Solutions that meet the design requirements are identified and the best solution is evaluated based on the performance measured while meeting the requirements.

10.4.3.1 Requirements

- Small and portable
- Affordable
- Fast in processing
- Eco-friendly
- Accurate

Considering the aforementioned facts, IR sensor suits our model best.

10.4.4 Functional Decomposition

Main function: Sorting and counting coins

Sub functions are as follows:

1. Sorting coins
2. Counting coins

10.4.4.1 Sorting Coins

Figure 10.12 shows the conceptual design used to sort coins. This model consists of holes of different diameters. Each hole corresponds to a particular coin, i.e., 1-rupee, 2-rupee, 5-rupee,

FIGURE 10.12 Coins are sorted (photographer courtesy: Tharun KM).

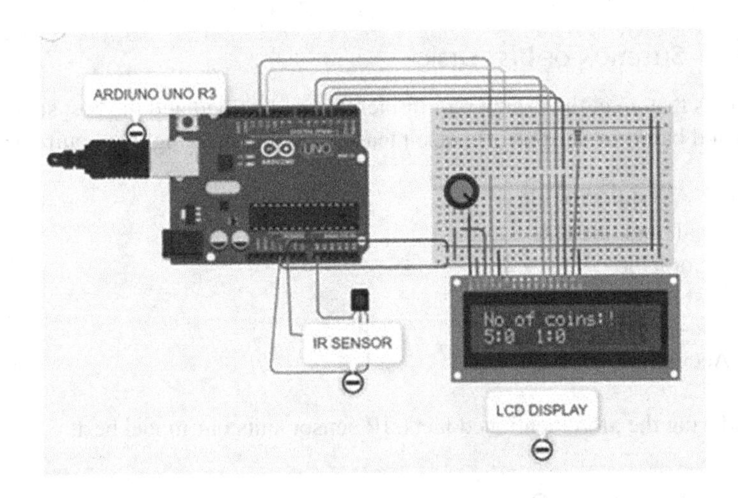

FIGURE 10.13 The hardware used in the system to count (photographer courtesy: Tharun KM).

and 10-rupee coin. These holes are placed in a slider. Once the coin slides through the slider, it automatically falls into its corresponding slot and gets sorted out.

10.4.4.2 Counting Coins

Counting coins is done through Arduino program and IR sensor. Once the coin cuts the IR sensor, a count is added and the value is displayed in the LCD display. The prototype is initially developed virtually using Tinkercad simulation software and is shown in Figure 10.13.

10.4.5 Development of Low-Cost Prototypes

Mechanical design to sort the coins is made. Figures 10.14 to 10.16 are the different views of the same mechanical model of the sorting tool.

Cardboard model is created, and the top and front views of the model are given in Figures 10.14 and 10.15, respectively. Figure 10.16 provides the side view.

The circuit board connection is shown in Figure 10.17, and the electrical prototype of the proposed coin counter is shown in Figure 10.18.

FIGURE 10.14 Top view of the prototype (photographer courtesy: Tharun KM).

FIGURE 10.15 Front view of the prototype (photographer courtesy: Tharun KM).

FIGURE 10.16 Side view of the prototype (photographer courtesy: Tharun KM).

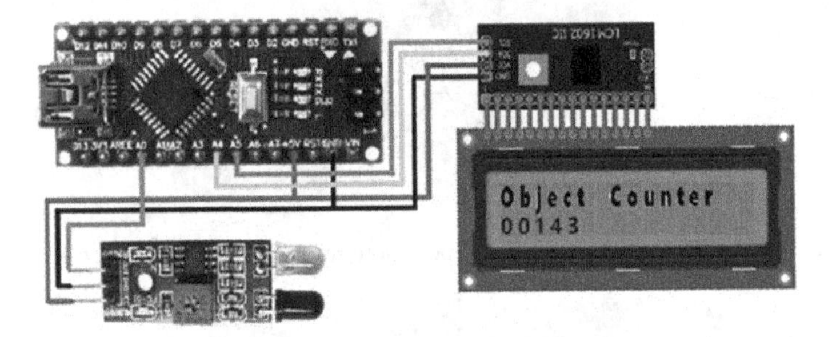

FIGURE 10.17 Circuit board connection (photographer courtesy: Tharun KM).

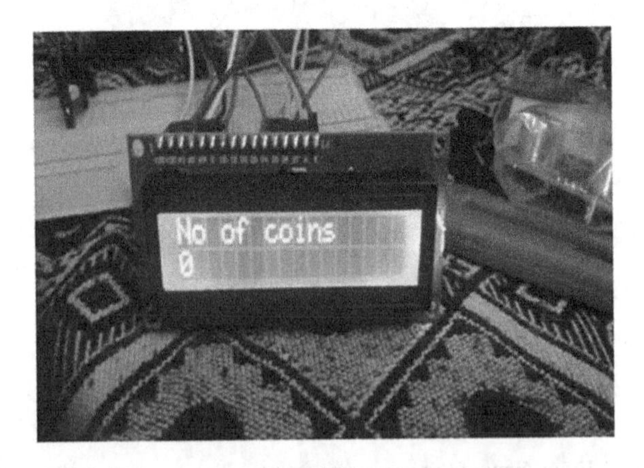

FIGURE 10.18 Monitoring using LCD (photographer courtesy: Tharun KM).

TABLE 10.5
Comparison of the Proposed Product with Market Product

Criteria	Existing Product	Proposed Product
Cost	Very high	Very much affordable and costing around INR 1,500
Maintenance	It is very much complicated and hard to maintain	Simple and easy to maintain
Size	Compact	Comparatively large
Coins' flow	Coins may get stuck inside	Flow of coins is smooth, and coins will not get stuck due to their large size.
Effectiveness	Effectively counts and sorts coins	Effectively counts and sorts coins
Preferences	Less preferred due to its high cost.	Highly preferred because of its affordable cost.

10.4.6 Comparison with the Existing Product

The proposed model is compared with the existing product as discussed in Section 10.3.2. Table 10.5 provides the comparison between the two.

10.4.7 Refinement of Design Specification on Users' Feedback

- Size of the product has been decreased to make it portable.
- LCD display is added to show the total value of coins.
- Batteries are added to make it rechargeable.

10.5 CONCLUSION

The aim of the project is to develop a product that can count and sort coins in a simple manner, which is affordable.

Coins are used by different people, who are in need to sort and count coins. There are products available in the market, but their price is very high and hence out of the reach for normal people in India.

When the proposed solution is implemented, the situation of dealing with coins will become much easier. It will decrease the investment in time and man power in counting and sorting of coins. This project will satisfy the requirements of the stakeholders at a reasonable cost. We will make sure that our project reaches the last man of the society.

10.5.1 Description of Deliverables

Our project will make coin counting and sorting coins much easier and time efficient than manual counting. Our product will be affordable so that everyone can make use of it.

10.5.2 Scope for the Improvement

Our product finds it difficult when it encounters same diameter coins to separate. We are working on it to find an alternative to deal with it. Our product processes coins one by one, and we are looking to improve it so that a bunch of coins can be processed at one.

10.5.3 Learning Out of This Project

- The importance of team work is realized in creating a quality product.
- Literature review of automatic counting and sorting coins is done.
- Steps involved in creating design and developing into products to compete in the market is studied.

10.5.4 Expected Time to Obtain the Real Solution to the Community Problem

The expected time to obtain the real solution to the community problem can take 3 to 4 months and is currently in progress.

REFERENCES

1. Guan, Goh Chin. "Design and implementation of an automatic coin sorting and counting machine." *University of Malaysia Pahang* (2015 June).
2. Kavale, Anupa, Shraddha Shukla, and Prachi Bramhe. "Coin counting and sorting machine." 9th International Conference on Emerging Trends in Engineering and Technology-Signal and Information Processing (ICETET-SIP-19). IEEE, 2019.
3. https://en.wikipedia.org/wiki/Currency-counting_machine
4. Yang, Yahan, and Xu Si. "Design of coin sorter counter based on MCU." *AIP Conference Proceedings*, Vol. 1955, No. 1 (2018). AIP Publishing LLC.
5. Feng, Bo-Yuan, et al. "Computer recognition and evaluation of coins." *Handbook of Pattern Recognition and Computer Vision* (2016), 141–158.
6. GONG, Yongzhen, et al. "Design and application of high efficiency coin sorting machine." *Machinery & Electronics* (2017), 7.
7. https://sdgs.un.org
8. www.freepatentsonline.com/9022841.html (Accessed on Jan 2021)
9. https://patents.google.com/patent/US8967361B2/en (Accessed on Jan 2021)
10. www.amazon.in (Accessed on Jan 2021)
11. www.sparkfun.com/products/10245 (Accessed on Jan 2021)
12. www.indiamart.com/proddetail/ir-sensor-21495675055.html (Accessed on Jan 2021)
13. www.novuslight.com/organic-cmos-image-sensor-market_N9398.html (Accessed on Jan 2021)

11 Development of Portable Food Heater Using Design Thinking Approach

C. Vigneswaran, M. Senthilkumar, R. Mohanraj, G. Madhan Mohan, Archana A., Brindha M., Kanmani M., Rajeshwari B. P., and Ruba Dharshini S.
PSG College of Technology, Coimbatore, India

CONTENTS

11.1 INTRODUCTION

Even though many individuals would love to consume cold meals on occasion, warm food is recommended. The cold food would cause intestinal sensitivities, sinus infections, and weariness; hence, it is best to avoid it. Therefore, the temperature of food or drink should be close to the body's normal temperature to

DOI: 10.1201/9781003189923-14

maintain a regular healthy metabolism. By staying indoors, it is easy to have access to warm food because of the availability of stationary equipment such as gas stoves, microwave ovens, induction cooktops, or even direct flame. But as soon as people step out, acquiring warm food can become a very big question mark.

The real-world challenge has aided in the development of a low-cost portable food heater that is ergonomically improved. The goal is to provide hot food wherever, at any time, and to improve human health and metabolism in the process.

As a result, a food heater will appeal to a wide range of people, especially working people and students. It is much desired to have a product that is both portable and practical. The food heater should allow individuals to quickly warm meals with a single click while still being portable. Direct current connected to a metal can conduct heat and can be used to warm food when placed on that metal. At the same time, it has to be portable; otherwise, it would require locating a plug-point. Heat can be produced through chemical processes, but it is not advisable to store food in the presence of chemicals.

As coils can be attached to a rechargeable battery and can be made portable, they would be harmless. The method has an advantage over all other heating methods. Copper, nichrome, and tungsten are heat-conducting materials that can be used to conduct heat to the body, and the body heat can be used to heat food.

A detailed literature review has been conducted by referring to multiple types of research on the topics of heat transfer modes, heat energy, medium transfer, and related material gleaned from various periodicals and publications. This study is supported by the following observations.

Murdani [2016] have studied the copper and nichrome wire safety limits for working at the specified current value for the indicated function. The authors also explain the reactions of copper and nichrome wires in fuse mechanisms. This analysis is carried out by altering the diameter of the wire for a certain current value, and the melting point is used to determine the safety limit. Peiravi and Javad [2020] have communicated a thorough understanding of heat transmission and its various kinds, as well as conduction, convection, and radiation in a body and its environment. They have explained heat transfer types, procedures, and concepts implied in calculations. This book describes the impacts and implications of such energy transfer in the environment. Mishra (2019) has discussed the challenges, primarily the loss of heat during delivery. Induction heating is used in this design, which is more efficient than battery-powered direct heating. The heating coil is made out of copper wire. They have also explained the product's development philosophy and approach followed by them.

Davis et al. [2020] have discussed the most important aspects of managing food materials and their heat transfer characteristics. The numerical approaches paved the path for the use of heat transfer ideas in the food business. In food engineering, this entails the analysis of finalizing heat transmission models. Finite differences, finite volume, and finite elements were among the key numerical approaches. The applications are also from the food engineering literature, as well as the principles of computational fluid dynamics for addressing fluid flows with heat transfer. Moitsheki and Atish [2011] discussed the phase instructions for simulating heat transfer models. The study describes the mode of heat transfer to employ a cooling system by emitting the heat created to the environment. In a two-dimensional system, this also computes the heat flux and its distribution. Dong and Bo Qin Gu. [2010] have simulated the model using ABAQUS and FLUENT to provide a complete examination of a solid–liquid coupling system and the heat transfer involved in the system. The efficiency of materials in emitting heat through finned radiators is also compared.

The literature survey clearly shows the importance of deriving the safety limit under the given criteria. The study of the properties of food and the heat models for food applications plays a vital role in this task. The methods for progressing the product's development also include heat transfer computation, methodologies, temperature, voltage, and current.

11.2 METHODS

The design thinking process is a five-step process (Brown [2008]): Problem identification, defining a suitable methodology, obtaining a frugal solution, innovation and implementation, and prototype development and testing.

11.2.1 Problem Identification

Hot food kills microbes and bacteria in it. In practice, an oven or direct flame is used to heat the food. But it is difficult to find a provision to warm the food in public or at workplaces. Though there are products that provide the function of warming the food, they are expensive, wired, boxes, space-consuming, and irreversible products.

The objective is to raise the temperature of the meals at the room temperature to an elevated temperature at a low cost. By doing this, the need to search for an expensive oven shall be minimized by providing a click on the lunch box to heat the food anytime and anywhere. The product has to be highly portable and should be able to be carried like a lunch bag.

11.2.2 Defining a Suitable Methodology

The customers' desires and expectations were observed through a survey. In connection to that, a suitable methodology had been framed for the identified problem. Hence, it was decided to formulate the objectives followed by generation of ideas, transforming ideas into conceptual sketches, generating CAD models by selecting a suitable material, calculating the power requirement, analyzing the simulated results, estimating costs, and finally producing the functional prototype and testing. A detailed methodology flow chart is shown in Figure 11.1.

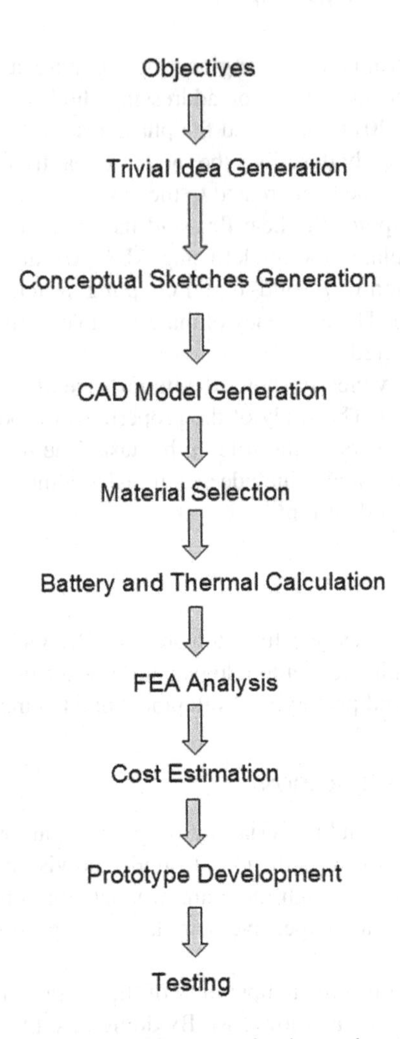

FIGURE 11.1 Detailed methodology flowchart for the product development.

11.3 OBTAINING A FRUGAL SOLUTION

The communication of conceptual ideas to obtain the required frugal solution would be the biggest challenge for a product designer (Bossert [2021]). Hence, developing conceptual sketches would solve the said purpose.

11.3.1 Generation of Conceptual Sketches

Among the developed conceptual sketches, the conceptual sketch of the portable food heater model is illustrated in Figure 11.2. The nichrome coil is wounded across the three faces of the container in the top view. A switch connects the coil to the battery. When the coil is heated, the heat is transferred to the container

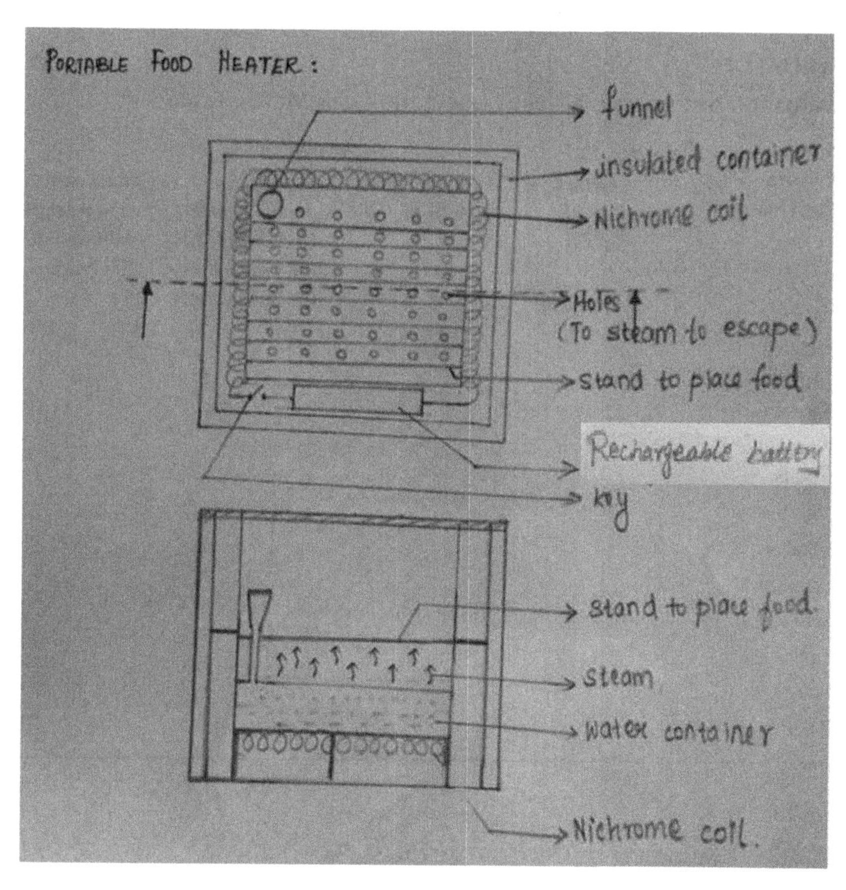

FIGURE 11.2 A pencil sketch of the conceptual design of the portable food heater.

walls. Furthermore, heat is transmitted to the meal by convection. The coil is wound on the bottom surface of the container in the front view. The generated heat is then transferred to the water above the coil winding. Water is poured into the water compartment using a conical funnel. The meal is warmed by the steam that escapes from the water. As a result, all three heat transfer modes are involved to heat the food.

After finalizing the conceptual idea, the said procedure is followed to generate different conceptual ideas for the base body. The elliptical body has been chosen as one of the generated conceptual ideas to maximize heat conduction and avoid hoop stress. The other ideas are given as follows: Bi-metallic strip was employed to control heat, similar to iron boxes. Manually assembling and disassembling the stand and water compartment would facilitate the ease of cleaning the container. Additionally, the rectangular body would help mount the battery at one face. Time and temperature adjustment of the product are provided to adjust and monitor the temperature against time. In this case, the said details would ensure a user-friendly product.

TABLE 11.1
Evaluation of Conceptual Ideas Using the Pugh Matrix Table

Selection Criteria Pugh Matrix	Datum	Conceptual Ideas			
		Elliptical Body	Functioning with Bi-Metallic Strip	Manual Assemble and Disassemble of the Product (Rectangle Body)	Time and Temperature Adjustment in the Product
Ease of handling	0	+	–	–	+
Ease of use	0	+	+	+	+
Portability	0	+	–	+	0
Ease of manufacture	0	–	–	+	–
Durability	0	0	–	+	–
Economical	0	–	+	+	–
Sum of +'s	0	3	2	5	2
Sum of -'s	0	2	4	1	3
Sum of 0's	0	1	0	0	1
Rank		2	4	1	3

11.3.2 Evaluation of Conceptual Ideas

As seen in the Table 11.1, the Pugh Matrix concept [8] was implied to evaluate the generated conceptual ideas. The best suited conceptual design has been chosen using selection criteria best-suited with the help of the Pugh Matrix.

The important parameters like the ease of manufacturability, ease of handling and use, portability, and durability were considered for evaluation. As per the ranking obtained from the Pugh matrix, the rectangular body with manual assembly and disassembling parts has been selected as the final design.

11.3.3 CAD Modeling

The inner view of the portable food heater modeled using PTC Creo 6.0 software is shown in Figure 11.3. The sectioned view provided in Figure 11.4 facilitates a better understanding of the selected conceptual idea.

11.3.4 Material Selection

Material selection is one of the crucial steps in product development. The proper and scientific method of material selection ensures product durability and

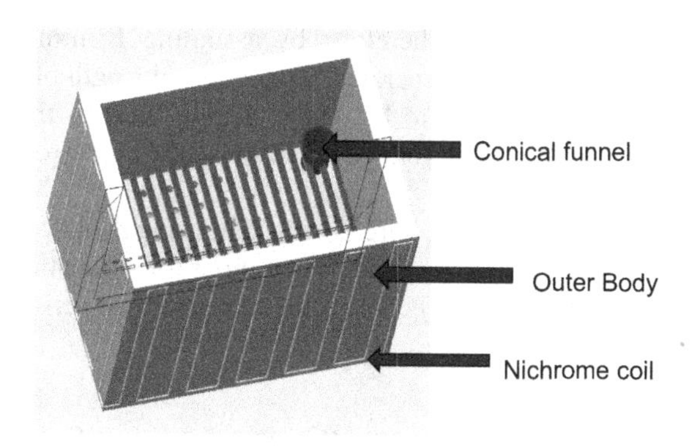

FIGURE 11.3 Creo model of portable food heater.

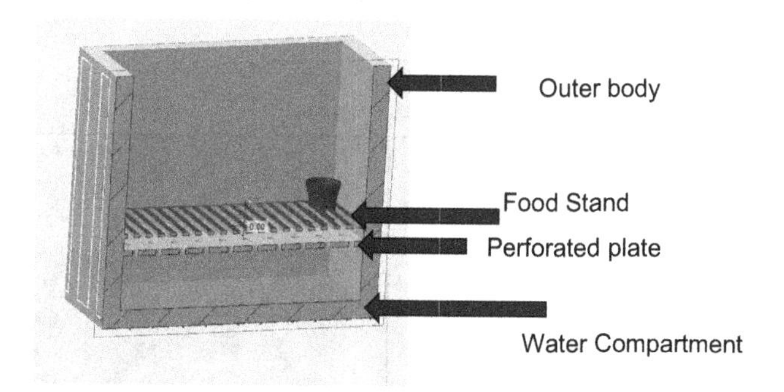

FIGURE 11.4 Sectional view of the portable food heater.

customer satisfaction. In this work, the Ashby chart (Ashby and Jones [2012]) has been employed to derive the suitable material for the outer body. The following factors were considered.

a. Objective: Maximum strength and maximum thermal conductivity
b. Constraint: Minimum weight
c. Candidate materials identified: Al alloys, Zn alloys, and Cu alloys

11.3.5 Power Supply Requirements

Based on the material selected, the following suitable power supply requirements have been identified with: the mass of the food heated is considered to be 0.45 kg. The ambient temperature is 30°C. The specific heat of food is taken as 3 kJ kg^{-1} °C^{-1}. The volume of the container to store food material is 2,560 cm^3. Overall wire resistance of the nichrome material is 8.106 Ω. The target temperature is assumed to be 212 °F (100 °C). The heat generated in the nichrome wire is calculated using Joule's Law of Heating. Therefore, by assuming 15 minutes of heating time, 10.3 kJ would be the heat generated. The required length of nichrome wire for sufficient heating was found to be 0.965 m (3.17 foot) as the resistance of nichrome is 2.56 Ω per food. The voltage required for the heat circuit is calculated to be 10.538 volts.

The energy requirement for the wire is arrived at using the law of conservation of energy. The arrived values were given as an input for simulation. The simulation is made in ANSYS and ABAQUS simulation software to verify the feasibility of the product mechanism.

11.3.6 Transient Thermal Analysis (Conduction and Convection)

The analysis is performed using ANSYS software in the transient thermal module as shown in Figure 11.5. This aims to analyze the temperature conducted and convected to the outer body and the inside body, respectively. The material

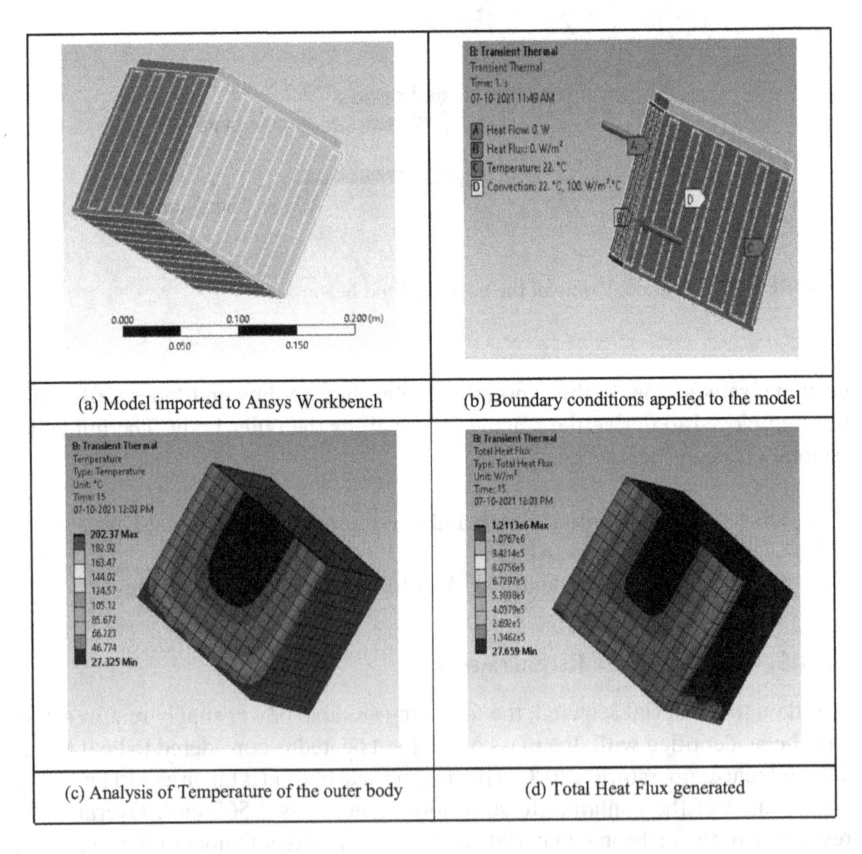

(a) Model imported to Ansys Workbench

(b) Boundary conditions applied to the model

(c) Analysis of Temperature of the outer body

(d) Total Heat Flux generated

FIGURE 11.5 Transient thermal analysis using ANSYS for conduction and convection.

of the coil is selected as nichrome from the Ashby Chart [9]. The temperature is given as input to the coil which was calculated manually. The dimension of the body is fixed as 16 × 16 × 10 cm^3, and the material is assigned as aluminum alloy. Boundary conditions for the analysis are temperature, heat flux, heat flow, and convection. The derived values from the calculation have been verified using transient thermal analysis. The results of the transient thermal analysis infer that the combined temperature attained through the conduction and convection process is sufficient to warm the food.

11.3.7 Radiation Thermal Analysis on Food Material

The analysis was performed to confirm that the given food material can be heated using the convected temperature obtained via calculation. Bread is taken as the reference food material for simulation. The properties of bread are assigned as follows:

1. Dimensions of the bread 50 × 50 × 100 mm
2. 2Heating duration 15 minutes
3. Emissivity 0.66 (no unit)
4. Input temperature for heating 110°C
5. Conductivity 0.1 mW/mm/K (= mJ/s/mm/K)
6. Specific heat capacity 2,650 mJ/T/K

As shown in Figure 11.6, the output is obtained in the form of gradient temperature of the bread.

11.3.8 Structural Analyses

The structural analysis of the outer body has also been performed to ensure the safety of the developed structure. Among the structures shown in Figure 11.7,

FIGURE 11.6 Gradient temperature of the bread sample during heating with a heat flow rate in the bread.

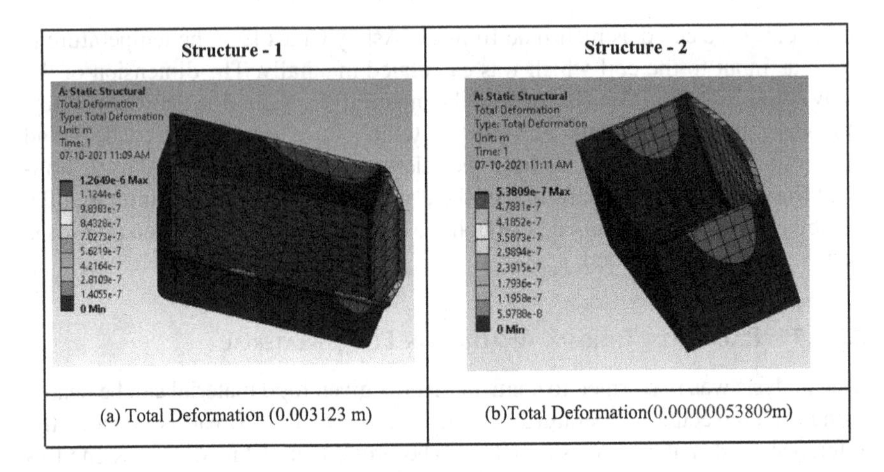

FIGURE 11.7 Structural analysis of the outer body.

TABLE 11.2
Structural Analysis Results of Structures 1 and 2

	Total Deformation	Equivalent Stress	Normal Stress
Structure 1	3.1×10^{-7} m	0.72 MPa	0.33 MPa
Structure 2	5.3×10^{-12} m	0.13 MPa	0.098 MPa

structures 1 and 2 were considered for structural analyses. The total deformation results obtained were tabulated in Table 11.2.

11.4 INNOVATION AND IMPLEMENTATION

The fourth step is to implement the results obtained through modeling and analysis. By comparing base body, total deformation, equivalent stress, and normal stress, it has been decided to select structure 2 as it is comparatively better than structure 1. The yield strength of the aluminum alloy is much larger compared to both the equivalent and normal stresses of the container. Hence, the clamping of the battery to one side of the container would not produce any major elastic or plastic deformational changes to the container. Additionally, the availability of the lid component along with structure 2 is an added advantage. The list of components with costs incurred for the functional prototype is shown in Table 11.3.

TABLE 11.3
List of Components and Costs Incurred

Sl. No	Component	Quantity	Cost (in Rs)
1	Sealed lead acid power battery Specifications: 12 V, 1.3 Ah Dimensions: 97 × 43 × 52 mm	1	600
2	Bread mold Dimensions: 20.32 × 10.16 × 10.16 cm	1	200
3	Nichrome wire Specifications: 26 gauge, 0.46 mm diameter	1	60
4	Sheet metal Dimensions: 10 × 6 inches Thickness: 0.4 mm	1	75
5	Natural rubber sheet Dimensions: 40 × 5 inches Thickness: 2 mm	1	100
6	Electrical switch	1	20
7	Connector clips	4	68
8	Heat string (1 m)	1	10
9	Connecting wire (1 m)	1	10

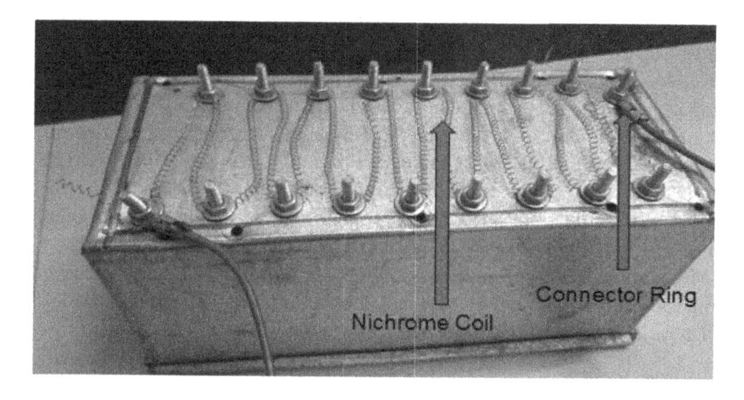

FIGURE 11.8 Fastening the coil to structure 2.

11.5 PROTOTYPE DEVELOPMENT AND TESTING

The functional prototype has been built. After constructing the functional model, the testing up of the prototype has been done. A water-filled aluminum foil container is placed inside the prototype. As illustrated in Figure 11.8, the prototype model is connected to the battery and tested. The lid of the developed model took nearly 20 minutes to feel the heat. Small air bubbles were seen in the aluminum foil container, indicating that the temperature of the water has been raised.

11.6 CONCLUSION

The material for the product development was carefully chosen scientifically using Ashby tables. The essential calculations for the chosen material were completed manually, and later the developed innovation was subjected to modeling and analysis and the obtained results were simulated and verified. The analysis and results of the developed product fared exceptionally well. As a result, the unique product would reach a wide range of customers and become a useful tool. The innovative product was prototyped using frugal and practical design thinking approach.

REFERENCES

Ashby, Michael F., and David R. H. Jones. *Engineering Materials 1: An Introduction to Properties, Applications and Design*, vol. 1. Elsevier, 2012.

Bossert, James L. *Quality Function Deployment: A Practitioner's Approach.* CRC Press, 2021.

Brown, Tim. "Design thinking." *Harvard Business Review* 86, no. 6 (2008): 84.

Davis, Yasmin, Nora Drewes, Thomas Lomax, and Sarah Norman. "A1_1 Undercooked in Overcooked?" *Physics Special Topics* 19, no. 1 (2020).

Dong, Jin Shan, and Bo Qin Gu. "Coupling analysis of heat transfer in finned radiator based on numerical simulation codes ABAQUS and FLUENT." *Advanced Materials Research* 118 (2010): 635–639. Trans Tech Publications Ltd.

Mishra, Ketaki. "From the determination of thermal properties of fibers to multiscale modeling of heat transfer in composites." Ph.D. diss., Nantes, 2019.

Moitsheki, Raseelo J., and Atish Rowjee. "Steady heat transfer through a two-dimensional rectangular straight fin." *Mathematical Problems in Engineering* 2011 (2011): Article ID 826819, 13 pages. https://doi.org/10.1155/2011/826819

Murdani, E. "Characterization of copper and nichrome wires for safety fuse." *Journal of Physics: Conference Series* 776, no. 1 (2016): 012099. IOP Publishing.

Peiravi, Mohammad Mohsen, and Javad Alinejad. "Hybrid conduction, convection and radiation heat transfer simulation in a channel with a rectangular cylinder." *Journal of Thermal Analysis and Calorimetry* 140, no. 6 (2020): 2733–2747.

Index

N

O

P

R

S

T

U

V

Printed in the United States
by Baker & Taylor Publisher Services